Phytochemicals in Agriculture and Food

Edited by Marcos Soto-Hernández,
Eva Aguirre-Hernández
and Mariana Palma-Tenango

Published in London, United Kingdom

Phytochemicals in Agriculture and Food
http://dx.doi.org/10.5772/intechopen.104363
Edited by Marcos Soto-Hernández, Eva Aguirre-Hernández and Mariana Palma-Tenango

Contributors
Abiola M. Ayodele-Asowata, Aisha Idris Ali, Asiya Mohammad, Babawale Peter Olatunji, Eva Aguirre-Hernández, Ezekiel Olumoye Oyetunji, Fatima Idris Ali, Herve Kuietche Mube, Hilal Punoo, Iqra Qureshi, Marcos Soto-Hernández, Mariana Palma-Tenango, Munir Abba Dandago, Samkeliso Takaidza, Serge Cyrille Houketchang Ndomou

First published in London, United Kingdom, 2024 by IntechOpen
IntechOpen is the global imprint of INTECHOPEN LIMITED, registered in England and Wales, registration number: 11086078, 167-169 Great Portland Street, London, W1W 5PF, United Kingdom

British Library Cataloguing-in-Publication Data
A catalogue record for this book is available from the British Library

Additional hard and PDF copies can be obtained from orders@intechopen.com

Phytochemicals in Agriculture and Food
Edited by Marcos Soto-Hernández, Eva Aguirre-Hernández and Mariana Palma-Tenango
p. cm.

This title is part of the Biochemistry Book Series, Volume 53
Topic: Chemical Biology
Series Editor: Andrei Surguchov
Topic Editors: Murat Şentürk and Deniz Ekinci

Print ISBN 978-1-83768-682-7
Online ISBN 978-1-83768-683-4
eBook (PDF) ISBN 978-1-83768-684-1
ISSN 2632-0983

For EU product safety concerns:
IN TECH d.o.o., Prolaz Marije Krucifikse Kozulić 3, 51000 Rijeka, Croatia,
info@intechopen.com or visit our website at intechopen.com.

IntechOpen Book Series

Biochemistry

Volume 53

Aims and Scope of the Series

Biochemistry, the study of chemical transformations occurring within living organisms, impacts all of the life sciences, from molecular crystallography and genetics, to ecology, medicine and population biology. Biochemistry studies macromolecules - proteins, nucleic acids, carbohydrates and lipids –their building blocks, structures, functions and interactions. Much of biochemistry is devoted to enzymes, proteins that catalyze chemical reactions, enzyme structures, mechanisms of action and their roles within cells. Biochemistry also studies small signaling molecules, coenzymes, inhibitors, vitamins and hormones, which play roles in the life process. Biochemical experimentation, besides coopting the methods of classical chemistry, e.g., chromatography, adopted new techniques, e.g., X-ray diffraction, electron microscopy, NMR, radioisotopes, and developed sophisticated microbial genetic tools, e.g., auxotroph mutants and their revertants, fermentation, etc. More recently, biochemistry embraced the 'big data' omics systems. Initial biochemical studies have been exclusively analytic: dissecting, purifying and examining individual components of a biological system; in exemplary words of Efraim Racker, (1913 –1991) "Don't waste clean thinking on dirty enzymes." Today, however, biochemistry is becoming more agglomerative and comprehensive, setting out to integrate and describe fully a particular biological system. The 'big data' metabolomics can define the complement of small molecules, e.g., in a soil or biofilm sample; proteomics can distinguish all the proteins comprising e.g., serum; metagenomics can identify all the genes in a complex environment e.g., the bovine rumen.

This Biochemistry Series will address both the current research on biomolecules, and the emerging trends with great promise.

Meet the Series Editor

Andrei Surguchov, Ph.D., joined Baylor College of Medicine, Houston, TX, as a faculty member in 1992, where he studied the mechanisms of the genetic control of lipid metabolism. At the University of Utah, his research interests focused on cloning new genes encoding retinal proteins. He studied molecular and cellular mechanisms of neurodegenerative diseases and retinal degeneration at Washington University, St. Louis. Currently, his research focuses on the structure-function relationship of proteins involved in neurodegeneration and ocular diseases. Andrei Surguchov is an Editor-in-Chief at Biochemistry Research International and Associate Editor in several biomedical journals.

Meet the Volume Editors

Dr. Marcos Soto Hernández is a pharmacist from the National Autonomous University of Mexico (UNAM). He completed his Ph.D. at the University of Wales, UK. He is currently a full professor at Colegio de Postgraduados, Mexico, where he researches the phytochemistry and bioactivity of natural products. He has established collaboration with research groups in the United Kingdom, the Netherlands, Spain, and México. He has received several awards locally and abroad. His main line of research is related to the bioguided isolation of secondary metabolites with importance in medicine and agriculture, as well as the potential of local aromatic plants.

Dr. Mariana Palma Tenango is an agricultural engineer from the Autonomous University of Chapingo, Mexico, with a Ph.D. in Plant Physiology from the Colegio de Postgraduados, Mexico. She is a teacher and researcher in the Faculty of Science, National Autonomous University of Mexico (UNAM). She is a reviewer and editor for several journals. She has participated in organizing congresses and symposiums in Mexico and is a supervisor of master's and doctoral students. Her research interests include phytochemistry, aromatic, and medicinal plants.

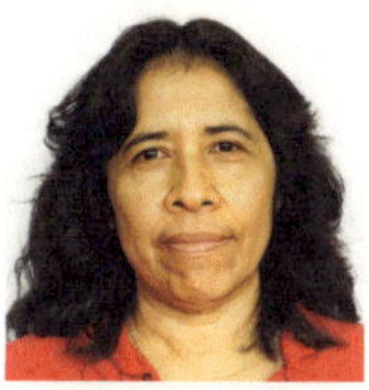

Dr. Eva Aguirre Hernández obtained a bachelor´s degree in biology from the National Autonomous University of Mexico (UNAM). She obtained her master's and doctorate degrees in science from the Graduate College, Montecillo campus, Texcoco, State of Mexico. She teaches courses in organic chemistry and plant biology as well as workshops in phytochemicals and pharmacology. Her research is related to natural products. She supervises undergraduate and postgraduate students and is a member of the tutorial committee of master's and doctorate postgraduate programs in biological sciences. She is a journal reviewer and has contributed to the dissemination of research at national and international conferences. Dr. Hernández is responsible for the laboratory of natural products of the Faculty of Sciences, UNAM.

Contents

Preface

Plants have developed numerous strategies to cope with the diverse biotic and abiotic factors existing in the ecosystems they inhabit. Among them is the production of a great diversity of secondary metabolites. About 200,000 compounds are recognized, which evolved in response or interaction with ecological aspects. The generation of chemical compounds by plants and their release into the environment can influence other organisms in different ways, for example, they can influence development and growth, health, behavior, and so on. This synthesis of natural products by plants allows the existence of particular biotic relationships, such as herbivory, allelopathy, pollination, and symbiosis, among others. In recent years, several phytochemicals present in various agricultural species have been discovered, providing added value to various fruits, seeds, grains, and more. The presence of these functional and nutraceutical compounds has motivated their research from different points of view such as their standardization, extraction methods, yield with respect to biomass, harvest dates, and their relationship with different processes.

Dr. Marcos Soto-Hernández
Full Professor,
Department of Botany,
Postgraduate College,
Estado de México, México

Dra. Mariana Palma-Tenango
Professor,
Faculty of Sciences,
The National Autonomous University of Mexico,
Mexico City, Mexico

Dra. Eva Aguirre Hernández
Full Professor,
Faculty of Sciences,
The National Autonomous University of Mexico,
Mexico City, Mexico

Chapter 1

Introductory Chapter: Phytochemicals in Foods

Marcos Soto-Hernández, Mariana Palma-Tenango and Eva Aguirre-Hernández

1. Introduction

Plants have evolved multiple mechanisms to adapt to the diverse biological and environmental challenges of the ecosystems they inhabit. One of these mechanisms is the synthesis of a wide range of secondary metabolites. These molecules play a crucial role in the adaptation of plants to their environment, but they also represent a significant source of active pharmaceutical products [1].

Recently, various phytochemicals have been identified in several plants, which have enhanced the value of fruits, seeds, and cereals, among others [2]. These compounds with functional and nutraceutical properties have been the subject of research covering topics such as their standardization, extraction techniques, yield concerning biomass, optimal harvest time, and their interaction with different processes.

2. Difference between the concept of nutraceutical and functional food

The concepts of nutraceuticals and functional foods have gained particular importance in contemporary nutrition. Although sometimes the term is used synonymously, there is a fundamental difference between them. While nutraceuticals refer to natural substances, extracts, or foods that have medicinal or health properties, functional foods are those that, in addition to their basic nutritional value, have ingredients that offer additional health benefits beyond mere nourishment.

As research delves deeper into the potential benefits of certain compounds found in foods, there is a growing interest in how bioactive compounds can be used not only to improve overall health but also to treat or prevent diseases.

Nutraceuticals, as their name suggests, lie at the intersection of nutrition and pharmaceuticals [3]. They come in various forms, from isolated natural ingredients to specific dietary products intended for health. These compounds can be beneficial in specific doses and, in some cases, may have therapeutic effects similar to those of medications.

Given that these compounds come from natural sources, such as plants, there is vast potential to discover new nutraceuticals from plant species that have not yet been studied in depth. This has led to a resurgence in the exploration of biodiversity in search of plants with potential health benefits. For instance, anthocyanins and other flavonoids are considered important nutraceuticals primarily due to their antioxidant effects, which gives them a potential role in the prevention of various diseases associated with oxidative stress [4].

Functional foods are foods whose consumption can have associated health benefits. This is in addition to the basic nutritional properties that foods possess [5]. Classic examples include dairy products enriched with probiotics, cereals fortified with vitamins and minerals, or drinks with added antioxidants. When consumed as part of a regular diet, these foods can offer additional health benefits, such as reducing the risk of chronic diseases.

Through biological and ecological interactions, plants have developed a wide variety of compounds that are now being explored for their potential benefits to human health. Understanding how these compounds interact at the cellular and molecular levels has driven innovation in the food industry. The current challenge lies in how to maximize the extraction and preservation of these compounds. Factors such as seasonality, genetic variability, and growth conditions can influence the concentration and efficacy of nutraceuticals and functional compounds in foods.

Interdisciplinarity has become a crucial tool in this field. Collaboration between botanists, chemists, nutritionists, and other professionals is leading to significant advances in the identification and application of these compounds.

Plants continue to play their role in ecosystems, producing chemical compounds in response to ecological factors, and humanity benefits from discovering, studying, and applying these compounds in the pursuit of optimal health.

Author details

Marcos Soto-Hernández[1*], Mariana Palma-Tenango[2] and Eva Aguirre-Hernández[2]

1 Colegio de Postgraduados, Campus Montecillo, Estado de México, México

2 Faculty of Sciences, National Autonomous University of México, México

*Address all correspondence to: msoto@colpos.mx

References

[1] Bourgaud F, Gravot A, Milesi S, Gontier E. Production of plant secondary metabolites: A historical perspective. Plant Science. 2001;**161**:839-851. DOI: 10.1016/S0168-9452(01)00490-3

[2] Maqsood S, Adiamo O, Ahmad M, Mudgil P. Bioactive compounds from date fruit and seed as potential nutraceutical and functional food ingredients. Food Chemistry. 2020;**308**:125522. DOI: 10.1016/j.foodchem.2019.125522

[3] Keservani RK, Kesharwani RK, Vyas N, Jain S, Raghuvanshi R, Sharma AK. Nutraceutical and functional food as future food: A review. Der Pharmacia Lettre. 2010;**2**:106-116

[4] Anderson OM, Marham KRC. Biochemistry and Applications. 2006;**45**:406-407. ISBN: 9780849320217

[5] Gul K, Singh AK, Jabeen R. Nutraceuticals and functional foods: The foods for the future world. Critical Reviews in Food Science and Nutrition. 2016;**56**:2617-2627. DOI: 10.1080/10408398.2014.903384

Chapter 2

The Use of Plants as Phytobiotics: A New Challenge

Serge Cyrille Houketchang Ndomou and Herve Kuietche Mube

Abstract

The search for bioactive compounds of natural origin, also called phytobiotics, has become a major challenge for industrialists, farmers, and scientists alike. Phytobiotics are compounds known for their anti-inflammatory, antioxidant, anti-carcinogenic, immunomodulatory, hypolipidemic, detoxifying, flavoring, and digestive-stimulating properties. These beneficial effects of phytobiotics depend on the part of the plant used (bark, leaves, stem, roots, fruit, flower, seeds) or their extract. Regarding their classification, there are several types of active compounds derived from plants, also grouped under the name of secondary metabolites such as tannins, polyphenols, terpenes, saponins, flavonoids, alkaloids, cyanides, and glycosides. Concerning their role, phytobiotics are used as feed additives to improve growth performance, nutritional status, and biochemical parameters of humans and animals. They can also be used ethno-medically for the prophylaxis and curative treatment of diseases such as diabetes, obesity, kidney stones, insomnia, gout, hemorrhoids, acne, and eye problems.

Keywords: phytobiotics, growth performance, nutritional and medicinal benefits, antibiotic growth promoters, secondary metabolites

1. Introduction

Using antibiotic growth promoters (AGPs) in livestock has been banned in the European Union countries since 2006 [1]. These regulations were developed to limit the consequences associated with bacterial resistance to antibiotics, which constitute a risk to both animal and human health [2]. The emergence of bacterial resistance in farm animals such as poultry is linked to the continuous administration of antibiotics at levels between 5 and 50 ppm [3]. Moreover, some of these AGPs would be involved in the development of pathogenic bacteria due to the reduction of the commensal of the microbiota [4]. They can also be found in small concentrations in meat [5], or high concentrations in animal waste. On the other hand, it has been shown that the presence of antibiotic residues in waste used as agricultural fertilizers is harmful to the environment by altering the availability of soil nutrients due to changes in microflora and microfauna and leading to the development of AGP-resistant bacteria in the soil [6].

Faced with these findings, the use of alternatives to AGP in animal feed has developed since the end of the 1990s, marked by the development of a large number of these products. Among these, the most common are prebiotics (substrates for the growth of certain bacteria of the digestive microbiota and indigestible by the host animal), probiotics (living microorganisms), certain organic fatty acids and enzymes, and natural phytobiotics from plants [7].

Constituted of several molecules with various properties, phytobiotics are promising products. For example, it has been shown that the antibacterial activities of phytobiotics are associated with compounds of various kinds that act on microorganisms by targeting different mechanisms of their physiology essential to their survival and thus limiting the risk of resistance development [8]. The composition of active substances in phytobiotics depends considerably on certain factors such as the part of the plant used, the climate, the state of maturity of the plant at harvest, and the nature of the soil, ... [9]. Moreover, it has been shown that the antioxidant properties of natural plant-derived phytobiotics are mainly associated with their content of phenolic compounds whose actions are identical to synthetic phenolic antioxidants [10].

However, ignorance of the specific and efficient conditions of the use of phytobiotics could contribute to understanding their differences in effectiveness. On the other hand, the nature and number of compounds used, the genetics of the animals, the composition of feeds, and the rearing conditions used could also explain the wide variability in the results reported in the efficacy of phytobiotics as promoters of growth in animals [7]. Thus, from these observations and analyses, it appears that many phytobiotics are full of molecules with various properties and capable of improving the growth performance and health of animals and even humans. Thus, their actions must be studied and valued, hence the objective of this study.

2. Generalities on phytobiotics

2.1 Definition

Phytobiotics come from plants and can have positive effects on the growth and health of animals due to their antibacterial, anti-inflammatory, and antioxidant properties [11]. They have also been defined as non-nutritive compounds and are therefore distinguished from the nutrients found in plants, such as vitamins and minerals. Phytobiotics consist of products of plant origin such as herbs, oleoresins, and essential oils; They can be added to the diet of commercial animals to increase their productivity by improving their feeding properties, promoting the production performance of animals and improving the quality of products derived from these animals. In addition, [8] have classified phytobiotics according to their origin and transformation, on this basis we can have essential oils, oleoresins, spices, and herbs.

2.2 Characteristics of phytobiotics

One of the most important criteria to take into account when choosing phytobiotics in animal husbandry is their general acceptability by the animals because if the organoleptic properties are not met, it is difficult for these phytobiotics to be accepted by the animals. In addition, among the biological activities of phytobiotics, some have a particular interest in making phytobiotics act on animal growth performance, antibacterial properties, stimulating digestion, anti-inflammatory, and antioxidant properties. Taking the latter case, the antioxidant properties of phytobiotics supplemented in feed could also improve the oxidative status of animals, which could have a positive impact on both animal health and the quality of their meat [12]. These functions can be performed by different families of compounds such as phenolic compounds, alkaloids, and terpenoids [13, 14].

By using phytobiotics in animal feed, the cost must not lead to an excessive increase in the price of the feed. Hence the importance of using by-products of other productions as the source of phytobiotics [7].

The ability of animals to perceive phytobiotics with odorous properties during feeding depends on the animal species, the molecule, and the portion of food considered which will prepare the digestive tract for intake, by stimulating digestive secretions and gastric motility [15, 16]. Thus, odorous compounds can therefore promote efficient digestion and so, improve the growth performance of animals.

2.3 Active ingredients of phytobiotics: plant secondary metabolites

Secondary metabolites are a group of various molecules involved in the adaptation of plants to their external environment. Indeed, there are more than 24,000 secondary metabolite structures involved in many mechanisms such as defense against herbivores and pathogens, regulation of symbiosis, control of seed germination, and chemical inhibition of competing plant species. Thus, secondary metabolites are integral to the interactions of species in plant and animal communities and the adaptation of plants to their environment. Some of the major plant secondary metabolites or phytochemicals found in plants include saponins, tannins, protease inhibitors, lectins, alkaloids, non-protein amino acids, and cyanogenic glycosides [17].

Types of secondary metabolites	Estimated number of structures
Nitrogen-containing metabolites	
Alkaloids	21,000
Lectins, peptides, and polypeptides	2000
Non-protein amino acids	700
Alkylamids	150
Amines	100
Glucosinolates	100
Cyanogenic glycosides	60
Metabolites without nitrogen	
Monoterpenes (C10)	2500
Sesquiterpenes (C15)	5000
Diterpenes (C20)	2500
Triterpenes, steroids, saponins (C30, C27)	5000
Tetraterpenes (C40)	500
Flavonoids, anthocyanins, catechins, tannins	5000
Phenylpropanoids, coumarin, lignans, lignin	2000
Polyacetylenes, fatty acids, waxes	1500
Polyketides	750

Table 1.
Plant secondary metabolites based on the presence or not of nitrogen [18].

Many criteria have to be considered for the classification of secondary metabolites including chemical structure, composition, solubility, and the biosynthetic pathway. According to [18], they can be grouped based on their composition by the presence or not of nitrogen (**Table 1**).

However, one of the most widely used criteria in the classification of secondary metabolites is the biosynthetic pathway where secondary metabolites of the plant could be grouped into three major groups namely: terpenes, alkaloids, and phenolic compounds [19].

2.3.1 Terpenes

In plants, terpenes constitute the largest group of secondary metabolites (SM) to which more than 40,000 different molecules are attributed. Structurally, they are unsaponifiable lipids since fatty acids are not involved in their formation. Since the basic structural unit that forms terpenes is isoprene, they are also known as isoprenoids. They can be classified based on their number of isoprene units (**Table 2**). For instance, hemiterpenes are the simplest class of terpenes, with a single isoprene unit and five carbons in its structure. Isoprene the best-known hemiterpene is a volatile product that results from photosynthetically active tissues. With two groups of isoprene, we have monoterpenes, sesquiterpenes with three units, diterpenes with four units, triterpenes with six units, tetraterpenes with eight units, and polyterpenes with more than 10 units [20]. Terpenes can be found in different plant parts such as flowers and fruits. These include lemon, ginger, mint, and eucalyptus. They act as defense molecules, toxic compounds, and food deterrents for insects. In some plants, terpenes function as a disperser by attracting pollinators [21].

2.3.2 Alkaloids

Like terpenes, alkaloids represent an important group of secondary metabolites comprising molecules primarily derived from vascular plants. It is noted that plants most often produce a complex mixture of alkaloids where a main constituent is mainly found. However, even if in a given plant their structures are slightly different,

Class	Number of isoprene units	Number of carbon atoms in the structure	Examples	Usages	Isolated from
Hemiterpene	1	5	Isovaleramide	Anticonvulsant	*Valeriana povonii*
Monoterpenes	2	10	Geraniol	Fragrance material	*Palmarose oil*
Sesquiterpenes	3	15	Farnesol	Source of perfume	*Citrus aurantium*
Diterpenes	4	20	Vitamin E	Antioxidant	*Corylus avellana L.*
Triterpenes	6	30	Squalene	UV protector	*Olive oil*
Tetraterpenes	8	40	Carotene	Antioxidant	*Rhodotorula glutinis*
Polyterpenes	>9	>40	Rubber	Restorative material (endodontics)	*Palaquium gutta*

Table 2.
Classes of terpenes according to the number of isoprene units [17].

the origin of their synthesis is common [22]. The number of alkaloids varies significantly between plants and parts of the same plant, we can also see the case where a plant does not contain any at all. Also, just like plants, they can be found in beings such as animals, fungi, and bacteria [23]. From a structural point of view, a nitrogen atom is generally found in alkaloids, and from a functional point of view, they are compounds that can be toxic and they commonly respond to precipitation reactions [24]. According to their biosynthetic origin, alkaloids are classified as true alkaloids, protoalkaloids, and pseudoalkaloids [25]. Although the presence of alkaloids is not vital to the plant, studies indicate that because of their deterrent ability and toxicity, they play a defensive role in the plant against insects and herbivores. While some alkaloids serve to protect the plant against certain predators (animals or microorganisms), others are used to fight other plant species in their habitat [26]. On the functional level, alkaloids have important physiological and toxicological properties exerted mainly at the level of the central nervous system (**Table 3**). Approximately 15,000 alkaloids have been isolated from plants to date. However, it is important to note that there is still a large amount of these compounds that have not been isolated from unidentified or no recorded higher plant species. Thus, from these observations and given the important medicinal, pharmacological, and therapeutic properties of alkaloids, more efforts must be made in the study of these compounds [27].

2.3.3 Phenolic compounds

The biosynthesis of phenolic compounds in plants is done from the two aromatic amino acids namely phenylalanine and tyrosine via the shikimic acid pathway. These compounds can have a simple or complex structure and the hydroxyl group (OH) of the aromatic ring is responsible for their antioxidant activity. Thus, polyhydrophenolics are those that contain more than two hydroxyl groups and phenolic compounds are those that contain more than one phenolic fraction [28]. Chemically, phenolic compounds are a very diverse group of secondary metabolites with phenol, the simplest representative of this class [29]. In the classification of phenolic compounds, one of the criteria to be considered is the number of carbon atoms present in the molecule, which makes it possible to have simple phenols, acid phenols, acetophenones,

Class	Name	Biological properties	Plant family
True alkaloids	Atropine	Anticholinergic drug.	Solanaceae
	Nicotine	A potent poison that at low doses is stimulating.	Solanaceae
	Morphine	Narcotic and anesthetic properties.	Papaveraceae
Protoalkaloids	Mescaline	Hallucinogen.	Cactaceae
	Hordenine	Stimulant of the central nervous system.	Cactaceae
	Ephedrine	Sympathetic nervous system stimulant.	Ephedraceae
Pseudoalkaloids	Aconitine	Highly poisonous.	Ranunculaceae
	Theobromine	Stimulating the central nervous system.	Malvaceae
	Coniine	Highly poisonous.	Apiaceae Sarraceniaceae

Table 3.
Some biologically relevant plant-derived alkaloids [17].

phenylacetic acids, hydroxycinnamic, coumarins, flavonoids, biflavonyls, benzophenones, xanthones, stilbenes, quinones, and betacyanins. In addition, compounds such as tannins, neolignans, lignans, and phlobaphenes can also be classified in the group of phenolic compounds (**Table 4**). Functionally, phenolic compounds can act as antioxidants [30], and they can also act as plant growth inhibitors [31]. Seeds can accumulate high numbers of phenols that act as filters so that oxygen does not reach the embryo and inhibit its germination [32]. In addition, phenolic compounds are responsible for the coloration and smell of fruits and thus make them appetizing for animals [33]. Plants also defend themselves against the attack of pathogens by synthesizing phytoalexins that are toxic to microorganisms and their presence prevents infections. Phenols also protect plants by generating bitter flavors or textures that are unpleasant for herbivores [34].

2.4 Good practice in the use of plant secondary metabolites

When studying secondary metabolites derived from plants, several steps must be taken into consideration, namely: extraction from plant sources, phytochemical screening of extracts for qualitative determination of metabolites present, purification of individual components and the elaboration of their chemical structures; the study of their biological activity (in vivo or in vitro assays), and the study of their toxicity or cytotoxicity on organisms or cells. However, to avoid possible chemical damage, the freshness of the plant samples must be maintained. Thus, the interval between the harvesting of plant species and the start of extraction should not exceed 3 hours, because due to the fragility of plant tissue, it deteriorates faster than dry tissue [35]. In addition, the most commonly found plant-drying processes are air-drying, microwave-drying, oven-drying, and lyophilization. But each of these methods has advantages and disadvantages that must be considered before their use [36, 37]. Another important point to consider during pre-treatment is the particle size of the plant material. Indeed, the smaller the particle size, the higher the contact area between the plant material and the solvent and, therefore, the more efficient the extraction of chemicals [38]. The various studies that exist on the effectiveness of

Skeleton structure	**Class**	**Characteristics**	**Examples**	**Isolated from**
C6	Simple phenolics.	Substituted phenols.	Hydroquinone	Bearberry plant
C6 – C1	Phenolics acids and related compounds.	A carboxyl group substituted on a phenol.	Galli acid	Banana
C6 – C2	Acetophenones and phenylacetic acids.	Are rarely found in nature.	2-hydroxyacetophenone	cocoa, coffee.
C6 – C3	Cinnamic acids, cinnamyl aldehydes, cinnamyl alcohols.	Are commonly found in plants as esters of quinic acids, shikimic acid, and tartaric acid or as sugar esters.	Sinapoyl choline	Seeds of *Arabidopsis thaliana*
C6 – C3	Coumarins, isocoumarins, and chromones.	They possess an oxygen heterocycle as part of the C3-unit.	Umbelliferone	Rutaceae and Apiaceae (Umbelliferae) families.

Skeleton structure	Class	Characteristics	Examples	Isolated from
C6-C3-C6 (C15) Flavonoids	Chalcones, aurones, dihydrochalcones.	Two benzene rings are linked together by a group of three carbons.	Butein	Dahlia and Coreopsis
	Flavones	Contain a ketone group, and an unsaturated C-C bond.	Kaempferol	Green leafy vegetables (spinach and kale), and herbs (dill, chives, and tarragon).
	Flavanones	Contain a ketone group.	Naringenin	Orange
	Flavanonols	Occur in association with tannins.	Taxifolin	onions, tamarind seeds.
	Anthocyanidins	The heterocycle is a pyrilium kation.	Cyanidin	Pomegranate, red cabbage, grapes.
	Anthocyanins	Are water-soluble glycosides of anthocyanidins.	Peonidin	Raw cranberries
C30	Biflavonyls	Are dimers of flavones or methylated derivatives.	Ginkgetin	*Ginko biloba* L.
C6-C1-C6	Benzophenones	Are aromatic ketones.	Benzophenones	Clusiaceae (Guttiferae)
	Xanthones	Are yellow pigments in flowers.	Xanthone	Mangosteen
C6-C2-C6	Stilbenes	Two benzene rings are linked together by a group of two carbons.	Resveratrol	Strawberry
C6, C10, C14	Quinones	Oxidizing agents	2,6-dimethoxy benzoquinone	*Rauvolfia vomitoria*, *Tibouchina pulchra.*
C18	Betacyanins	Are red pigments and contain nitrogen.	Betanidin	Cacti, carnations, amaranths, ice plants, beets.

Table 4.
Classes of phenolic compounds according to the number of carbons [17].

these compounds are only at the laboratory level, which is why it is still necessary to explore and evaluate their effectiveness at the greenhouse and field levels.

3. Properties of some plants as phytobiotics in livestock

Phytobiotics are used in poultry, but also in other rent, particularly in pigs and ruminants. The introduction of phytobiotics in animal feed was carried out by combining observations from "traditional" herbal medicine particularly important in certain regions of the world (China, Africa, South America), and rational phytotherapy based on scientific observations [7]. Over the past two decades, studies have shown that phytobiotics exert multiple effects such as anti-inflammatory, antimicrobial, antioxidant, and metabolic effects [39]. Phytobiotics are known to promote growth and improve meat and egg quality in poultry production [40]. Due to the gradual and continuous elimination of growth-promoting antibiotics due to biological resistance effects observed in animal production, the use of phytobiotics has expanded in animal husbandry to improve the growth performance of animals [41].

3.1 Antibacterial properties of phytobiotics

Medicinal herbs are used as a treatment for human diseases. The whole plant or part of the plant (leaves, roots, stem, flower, and fruits) is subjected to the extraction process for the derivation of the bioactive compound [42]. The inhibitory effect of bacterial growth has been demonstrated for many phytobiotics, and the intensity of their effect depends on both the target bacteria, the phytobiotics, and the dose used [7]. Indeed, several bioactive compounds from plants and fungi have demonstrated significant stimulatory properties of useful bacteria such as lactobacilli and bifidobacteria without however promoting the growth of pathogenic bacteria. Thus, the stimulation of these beneficial bacteria would contribute to the balance of the intestinal microflora and provide optimal conditions for effective protection against pathogenic microorganisms and an intact immune system in animals. In addition, in monogastric animals like pigs, spice-derived phytobiotics have been shown to stimulate appetite and endogenous enzyme secretion and exert coccidio-static or anthelmintic antimicrobial activities [43]. Researchers found that leaves of rosemary (*Rosmarinus officinalis*) at a concentration of 10 g/kg enhanced the immune system of Nile tilapia and increased their disease resistance against *Aeromonas hydrophila* [44]. In addition, many studies have revealed the potential of medicinal herbs for aquaculture uses; thus, researchers have developed new and improved approaches for antibacterial discoveries from plants. According to the literature, many medicinal herbs were reported to demonstrate antibacterial properties against *A. hydrophila*. For instance, *Murraya koenigii*, *Pandanus odoratissimus*, *Colocasia esculenta*, and *Euphorbia hirta* inhibited the growth of *A. hydrophila*. These medicinal herbs (**Table 5**) contain bioactive compounds, such as carbazole alkaloids, phenolic compounds, polypeptides, and alkaloids, that were responsible for antibacterial activities [45].

3.2 Antioxidant properties of phytobiotics

An antioxidant is any organic (or inorganic compound) presents at low concentrations compared with those of an oxidizable substrate, significantly delaying or preventing oxidation of that substrate [46]. The health benefits of the ingestion of phytobiotics are attributed to their antioxidant activity which has also been assessed from plant extracts. This activity is mainly associated with the presence of active compounds such as polyphenols. However, flavonoids would mainly act as buffers by capturing free radicals and transforming them into less reactive flavin radicals because their structure has delocalized electrons. On the other hand, quercetin can chelate transition metal ions such as iron or copper, thus preventing the synthesis of reactive oxygen species or free radicals [47]. Also, caffeic, chlorogenic, sinapic, ferulic, and *p*-coumaric acids have antioxidant activity by the inhibition of oxidation

Compound's family	Sub-family	Mechanism of action
Protein derivatives	Sulfur derivatives	Formation of ion channels in membranes via the formation of disulphide bridges.
	Lectins	Competitive inhibition of bacterial adhesion to their polysaccharide receptors.
	Alkaloids	Intersperse DNA.
Terpenes, terpenoids, and Saponins.	Terpenes, terpenoids.	Permeabilization of bacterial membranes.
	Saponins	Permeabilization of bacterial membranes.
Phenolic compounds	Simple phenols and phenolic acids.	Formation of complexes with many proteins, destruction of bacterial membranes, unavailability of certain substrates for bacteria.
	Quinones	Inactivation of membrane proteins.
	Flavones, flavanols, and flavonoids.	Binding to adhesins, binding to cell walls, inactivation of enzymes.
	Tannins	Binding to different proteins: adhesins, enzymes, substrate, cell wall, membrane. Formation of complexes with minerals.

Table 5.
Families of molecules with antibacterial activity and mechanisms of action (adapted from [7]).

of lipids and the elimination of reactive oxygen species, these effects are important to the plant defense [48]. According to [49], some polyphenolic compounds exhibit antioxidant effects through a variety of mechanisms that are regulated by antioxidant enzymes such as superoxide dismutase (SOD), catalase (CAT), and glutathione peroxidase, rather than the single mode of action of typical synthetic antioxidants (**Table 6**). In addition, the antioxidant activity of phytobiotics, especially phenolic acids, and flavonoids, is predominantly determined by the structure and electron delocalization over an aromatic nucleus [50].

According to their mechanism of action, there are primary antioxidants and secondary antioxidants. Primary antioxidants can neutralize free radicals by two mechanisms, either by the transfer of a hydrogen atom (most often labile hydrogen) (**Figure 1**) or by the transfer of a single electron (**Figure 2**). Primary antioxidants are very effective and are most often needed in small quantities and their high catalytic

Phenolic compounds	Pathological condition	Mechanism of actions
Catechins	Neurodegenerative diseases.	Enhance activity of SOD and catalase.
Ferulic acid	Diabetes, colon cancer.	Decrease lipid peroxidation and enhance the level of glutathione and antioxidant enzymes.
Quercetin, Procyanidins.	Oxidative stress in human astrocytoma U373 MG cell line.	Enhance the activity of antioxidant enzymes and expression of proteins.
Ellagic acid	Oxidative stress induced by oxidized lipids.	Enhance the activity of GSH and antioxidant enzymes.

Table 6.
Antioxidant properties of some phenolic compounds (adapted from [49]).

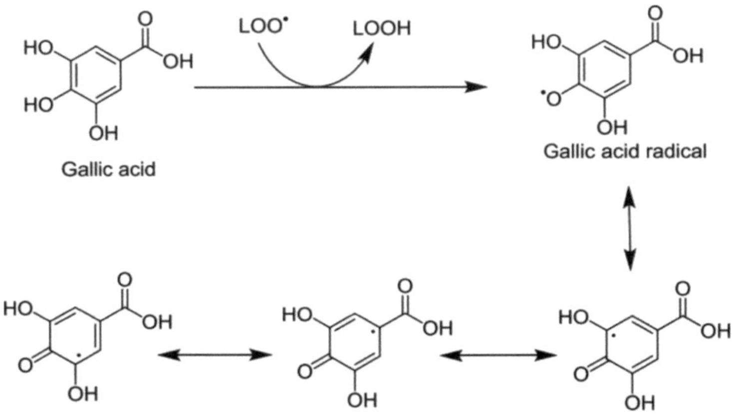

Figure 1.
The reaction of gallic acid with free radicals and its stabilization of gallic acid-free radical [49].

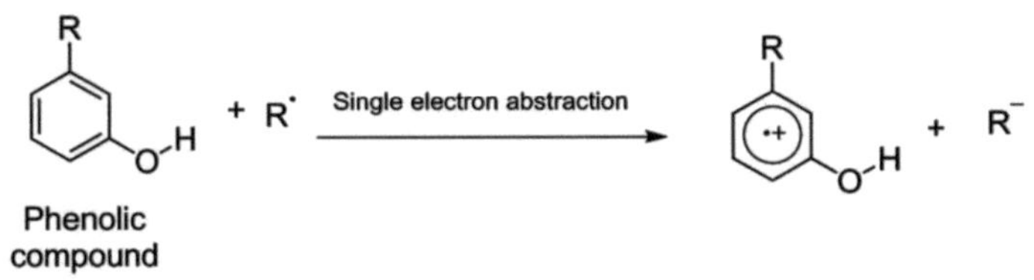

Figure 2.
Mechanism of single-electron abstraction reaction (SET) [49].

property is one of the reasons for their diversity in nature. Phenolic compounds are an example of these antioxidants and during their mechanism, they can be regenerated by resonance [49].

Secondary antioxidants can be distinguished by their mechanism of action. For example, ethylenediaminetetraacetic acid (EDTA) and citric acid can act as chelators of pro-oxidant metal ions (Fe^{2+} and Cu^{2+}). β-carotene can neutralize reactive species like singlet oxygen. Secondary antioxidants usually neutralize a free radical and are therefore easily depleted. Metal chelation can directly inhibit Fe^{+3} reduction, consequently reducing the formation of reactive OH-free radicals of the Fenton reaction [51] (**Figure 3**). The metal chelation depends on the reduction potential of the phenolic compounds; however, chelation is subject to certain conditions such as the metal ions do not attach to proteins or other chelator molecules.

More recently, a third class has been included called tertiary antioxidants. These antioxidants repair damaged biomolecules such as DNA or proteins. However, very little is known about their role in foods [49].

3.3 Growth performance properties of phytobiotics

Because of the increased risk of occurrence of antimicrobial resistance due to antibiotic growth promoters (AGPs), the use of phytobiotics as an alternative to AGPs has been extended to farm animals for improving their intestinal status and subsequently promoting growth [52]. So, numerous studies have reported the growth-promoting effects of phytobiotics in chickens, but their precise mechanism of action is yet to be elucidated [53]. However, a few reviews have suggested the

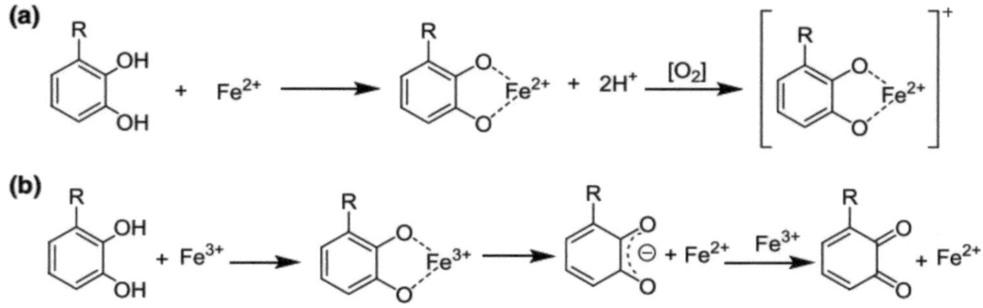

Figure 3.
Mechanism of metal chelation of phenolic antioxidants. (a) Coordination of Fe^{2+} by polyphenols and subsequent electron transfer reaction in the presence of oxygen-generating the Fe^{2+}-polyphenol complex; (b) coordination of Fe^{3+} by polyphenols, subsequent iron reduction and semiquinone formation, and reduction of Fe^{3+} to form a quinone species and Fe^{2+}. R = H, OH [51].

possible mechanisms by which phytobiotics lead to health benefits and growth promotion [54, 55]. According to [56], the effect of phytochemicals on growth performance elevation may associate with their antioxidant capacity or anti-inflammatory activity. Indeed, some natural phytochemicals represent a promising non-antibiotic tool for better intestinal health, nutrient digestibility, and general health status, thus leading to increase growth performance. In addition, it has been shown that supplementation with *Broussonetia papyrifera* leaf extract can increase the growth performance and antioxidant capacity of weaned piglets [57]. Supplementing rations with *Camellia sinensis* powders could be a good alternative to enhance the growth performance, carcass characteristics, and blood lipid profile of broilers [58]. However, regarding the biological activities of phytochemicals, 04 mechanisms (**Figure 4**) supporting the observations of physiological changes in animals (growth performance, carcass characteristics, and meat quality) have been proposed, namely: 1) Improvement of food status and consumption animal feed; 2) Modulator of ruminal fermentation; 3) Improved digestion and absorption of nutrients; and 4) Source of direct and indirect anabolic activity on target tissues [55].

3.4 Immune-activating properties of phytobiotics

By preventing and controlling infectious diseases in the animal population, phytobiotics derived from plants can be used as alternative products to minimize the need for antibiotics. For instance, the immune-activating properties of medicinal plants such as *Carthamus tinctorius*, *Taraxacum officinale*, and *Brassica juncea*, have been evaluated in vitro using avian lymphocytes and macrophages, and all their extracts stimulate innate immunity, inhibit tumor cell growth, and exert antioxidant effects in poultry [59]. In addition, cinnamaldehyde a constituent of cinnamon (*Cinnamomum cassia*) stimulated primary chicken spleen lymphocyte proliferation in vitro and activated macrophages to produce high nitric oxide (NO) [60]. Phytochemicals also exert their action through immunomodulatory effects such as the increased proliferation of immune cells, modulation of cytokines, and increased antibody titters [54]. In addition, it has been revealed that curcumin inhibits TLR4 (Toll-like receptors) and NOD (nucleotide-binding oligomerization) which are two primary targets of phytobiotics [61]. Two garlic metabolites namely propyl thiosulfinate (PTS) and propyl thiosulfinate oxide (PTSO) have been used in poultry feeding, and results revealed

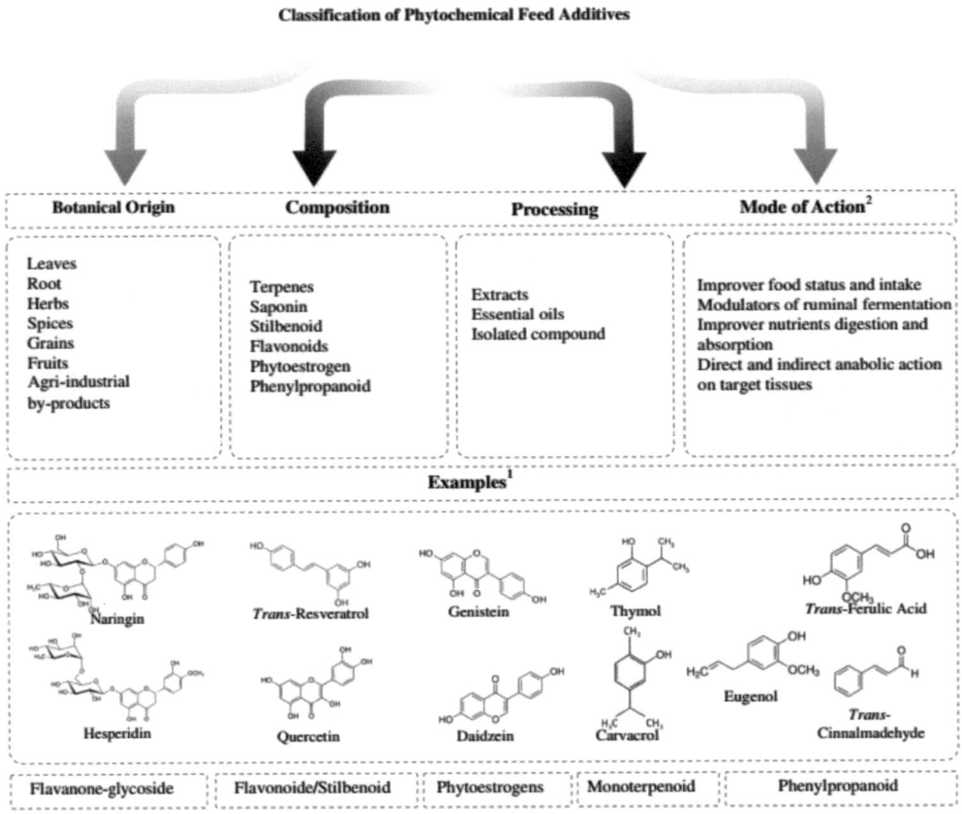

Figure 4.
The classification proposed and several examples of phytochemicals used as growth promoters' additives [55].

that supplementation of 10 mg/kg PTS/PTSO increased body weight gain and serum antibody titters against profilin, an immunogenic protein of *Eimeria*, and decreased fecal oocyst-excretion-in *Eimeria acervuline*-challenged chickens compared with chickens receiving a control diet. Also, adding PTS/PTSO in the broiler's diet, altered many genes related to innate immunity such as TLR3, TLR5, and NF-κB, and down-regulated expression of a cytokine such as IL-10 compared with the control diet [62]. Also, the combination of multiple phytochemicals exerts synergistic effects to reduce the negative consequences of enteric infections. A study revealed that the association of Curcuma/Capsicum/Lentinus-fed birds increased the levels of transcripts for IL-1β, IL-6, IL-15, and IFN-γ in gut lymphocytes compared with those fed the standard, Curcuma or Capsicum/Lentinus diet [63]. It can be observed that phytobiotics improve the intestinal inflammatory status and barrier functions, possibly via the inhibition of TLRs and subsequent activation of NF-κB, the activation of the xenobiotics detoxifying system, the reduction in pathogenic bacteria, and the Nrf2 pathway. This improvement in intestinal function subsequently prevents the translocation of pathogens and harmful constituents such as lipopolysaccharide (LPS) into the circulatory system and the induction of systemic inflammation via excess secretion of cytokines and glucocorticoids. The HPA (hypothalamic-pituitary-adrenal) axis controls glucocorticoid secretion, and excessive and long-term glucocorticoid secretion disrupts the protective function of the gut barrier and the gut microbiota [64]. These disturbances accelerate the production of cytokines, which in turn stimulate the HPA

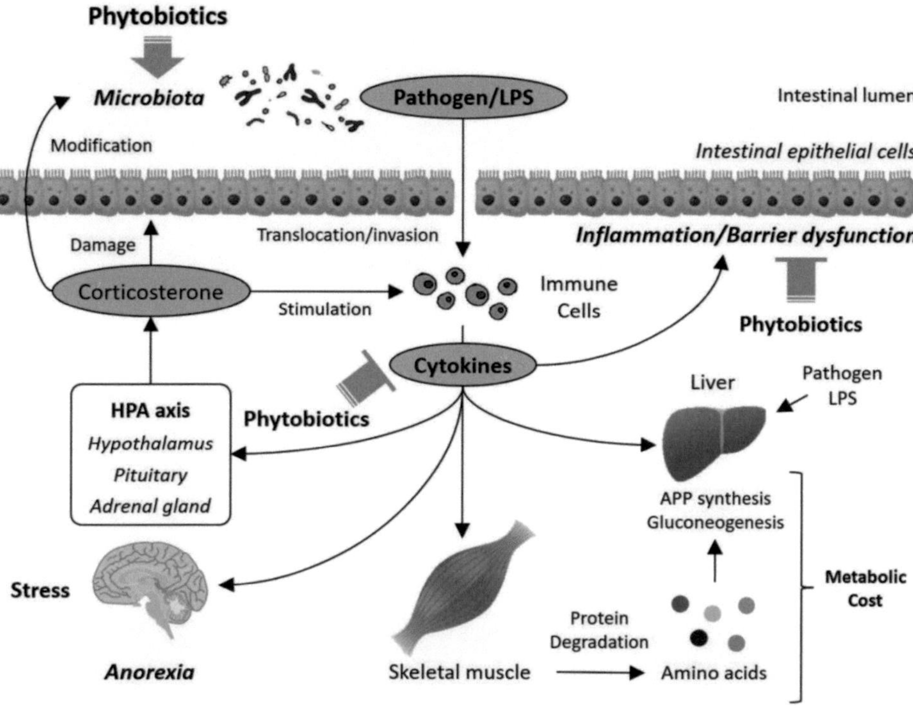

Figure 5.
Possible mechanisms of the mode of action in the beneficial effect of phytobiotics [53].

axis to secrete glucocorticoids. Thus, a vicious circle is initiated, which generates inflammatory and metabolic deregulations (**Figure 5**).

4. Conclusion

Increasing concerns about the increase of superbugs and limited development of new drugs for livestock necessitates the timely development of alternatives to antibiotic growth promoters. So, the trend in the use of phytobiotics in animal feed has increased during the last two decades. The health benefits and growth-promoting effects of phytobiotics may be dependent on several mechanisms based on their various biological activities. Also, many studies have been done using phytobiotics in livestock production. They have shown particularly the antimicrobial, antioxidant, anti-inflammatory, and growth-promoting effects of phytobiotics. The antioxidative function of phytobiotics can positively affect the stability of animal feed and increase animal products' quality and storage time. However, due to the contradictions in published results, further research and investigations are still necessary to elucidate various aspects such as the nutritional's aspect of phytobiotics.

Conflict of interest

The authors declare no conflict of interest.

Author details

Serge Cyrille Houketchang Ndomou[1,2]* and Herve Kuietche Mube[3]

1 Faculty of Agronomy and Agricultural Science, CRESA Forêt-Bois, University of Dschang, Yaoundé, Cameroon

2 Faculty of Science, Department of Biochemistry, Research Unit of Biochemistry, Medicinal Plants, Food Sciences and Nutrition, University of Dschang, Dschang, Cameroon

3 Faculty of Agronomy and Agricultural Sciences, Department of Animal Production, Research Unit in Animal Nutrition and Production (RUANP), University of Dschang, Dschang, Cameroon

*Address all correspondence to: sergendomou@gmail.com

References

[1] Commission Regulation (EC) No 429/2008 of 25 April 2008 on detailed rules for the implementation of Regulation (EC) No 1831/2003 of the European Parliament and of the Council as regards the preparation and the presentation of applications and the assessment and the authorisation of feed additives (OJ L 133, 22.5.2008, pp. 1-65)

[2] Diarra MS, Silversides FG, Diarrassouba F, Pritchard J, Masson L, Brousseau R, et al. Impact of feed supplementation with antimicrobial agents on growth performance of broiler chickens, Clostridium perfringens and enterococcus counts, and antibiotic resistance phenotypes and distribution of antimicrobial resistance determinants in Escherichia coli isolate. Applied and Environmental Microbiology. 2007;**73**:6566-6576

[3] Castanon JIR. History of the use of antibiotics as growth promoters in European poultry feeds. Poultry Science. 2007;**86**:2466-2471

[4] Asakura H, Tajima O, Watarai M, Shirahata T, Kurazono H, Makino S. Effects of rearing conditions on the colonization of salmonella enteritidis in the cecum of chicks. Journal of Veterinary Medical Science. 2001;**63**:1221-1224

[5] Weiss C, Conte A, Milandrib C, Scortichini G, Semprini P, Usberti R, et al. Veterinary drugs residue monitoring in Italian poultry: Current strategies and possible developments. Food Control. 2007;**18**:1068-1076

[6] Kumar K, Gupta SC, Chander Y, Singh AK. Antibiotic use in agriculture and its impact on the terrestrial environment. Advances in Agronomy. 2005;**87**:1-54

[7] Sarah G. Effets de phytobiotiques sur les performances de croissance et l'équilibre ou microbiote digestif du poulet de chair. ABES. 2011. p. 474. ID: 10670/1.dz7spd

[8] Windisch WM, Schedle K, Plitzner C, Kroismayr A. Use of phytogenic products as feed additives for swine and poultry. Journal of Animal Science. 2008;**86**:E140-E148

[9] Bakkali F, Averbeck S, Averbeck D, Idaomar MM. Biological effects of essential oils-a review. Food and Chemical Toxicology. 2008;**46**(2008):446-475

[10] Duong DN, Qin G, Harris JO, Hoang TH, Bansemer MS, Currie KL, et al. Effects of dietary grape seed extract, green tea extract, peanut extract, and vitamin C supplementation on metabolism and survival of greenlip abalone (Haliotis laevigata Donovan) cultured at high temperature. Aquaculture. 2016;**464**(2016):364-373. DOI: 10.1016/j.aquaculture.2016.07.011

[11] Zeng Z, Zhang S, Wang H, Piao X. Essential oil and aromatic plants as feed additives in non-ruminant nutrition: A review. Journal of Animal Science and Biotechnology. 2015;**6**:7

[12] Brenes A, Roura E. Essential oils in poultry nutrition: Main effects and modes of action. Animal Feed Science and Technology. 2010;**158**:1-14

[13] Gautam R, Jachak SM. Recent developments in anti-inflammatory natural products. Medicinal Research Reviews. 2009;**29**:767-820

[14] Serrano J, Puupponen-Pimia R, Dauer A, Aura AM, Saura-Calixto F. Tannins: Current knowledge of food

sources, intake, bioavailability and biological effects. Molecular Nutrition & Food Research. 2009;**53**:S310-S329

[15] Katschinski M. Nutritional implications of cephalic phase gastrointestinal responses. Appetite. 2000;**34**:189-196

[16] Teff K. Nutritional implications of the cephalic-phase reflexes: Endocrine responses. Appetite. 2000;**34**:206-213

[17] Mera IFG, Falconí DEG, Córdova VM. Secondary metabolites in plants: Main classes, phytochemical analysis and pharmacological activities. Bionatura. 2019;**4**:1000-1009. DOI: 10.21931/rb/2019.04.04.11

[18] Wink M. Evolutionary advantage and molecular modes of action of multi-component mixtures in phytomedicine. Current Drug Metabolism. 2008;**9**:996-1009

[19] Ávalos GA, Pérez-Urria CE. Metabolismo secundario de plantas. Reduca (Biología). Serie Fisiología Vegetal. 2009;**2**:119-145

[20] Taiz L, Zeiger E. Plant Physiology. 5th ed. Sunderland: Sinauer Associates Inc.; 2010. p. 782

[21] Tiago O, Nardino M, Carvalho IR, Follmann DN, Szareski VJ, Ferrari M, et al. Plant secondary metabolite and its dynamical systems of induction in response to environmental factors: A review. African Journal of Agricultural Research. 2017;**12**:71-84. DOI: 10.5897/AJAR2016.11677

[22] Roberts MF, Wink M. In: Roberts MF, editor. Alkaloids: Biochemistry, Ecology, and Medicinal Applications. New York, USA: Plenum Press; 1998. pp. 1-7 ISBN 978-1-4757-2905-4

[23] Bérdy J. Bioactive microbial metabolites. The Journal of Antibiotics. 2005;**58**:1-26

[24] Matsuura HN, Fett-Neto AG. Plant alkaloids: Main features, toxicity, and mechanisms of action. In: Gopalakrishnakone P, Carlini C, Ligabue-Braun R, editors. Plant Toxins. Toxinology. Dordrecht: Springer; 2017 ISBN 978-94-007-6463-7

[25] Aniszewski T. Alkaloids – Secrets of Life. Alkaloid Chemistry, Biological Significance, Applications and Ecological Role. Joensuu, Finland: Elsevier; 2007 Amsterdam • Boston • Heidelberg • London • New York • Oxford • Paris. San Diego • San Francisco • Singapore • Sydney • Tokyo

[26] Kartsev VG. Natural compounds in drug discovery. Biological activity and new trends in the chemistry of isoquinoline alkaloids. Medicinal Chemistry Research. 2004;**13**:325-336

[27] Azimova S, Marat Y. Natural Compounds-Alkaloids. New York: Springer Science Business Media; 2013

[28] Dimitrios B. Sources of natural phenolic antioxidants. Trends in Food Science & Technology. 2006;**17**(9):505-512. DOI: 10.1016/j.tifs.2006.04.004

[29] Velderrain-Rodríguez GR, Palafox-Carlos H, Wall-Medrano A, AyalaZavala JF, Chen C-YO, Robles-Sanchez M, et al. Phenolic compounds: Their journey after intake. Food & Function. 2014;**5**:189-197. DOI: 10.1039/c3fo60361j

[30] Martins N, Barros L, Ferreira IC. In vivo antioxidant activity of phenolic compounds: Facts and gaps. Trends in Food Science & Technology. 2016;**48**:1-12. DOI: 10.1016/j.tifs.2015.11.008

[31] Sudha Rani C, Sudhakar C. Effect of plant growth retardants on growth, yield and economics of kharif pigeonpea (Cajanus cajan. L). International Journal

of Chemical Studies. 2018;**6**:1495-1498 P-ISSN: 2349-8528. E-ISSN: 2321-4902

[32] Stanisław W, Chrzanowski S, Karamać M, Król A, Badowiec A, Mostek A, et al. Analysis of phenolic compounds and antioxidant abilities of extracts from germinating Vitis californica seeds submitted to cold stress conditions and recovery after the stress. International Journal of Molecular Sciences. 2014;**15**:16211-16225. DOI: 10.3390/ijms150916211

[33] Kim V, Burke RJ, Styler SA, Jackson DA, Melin AD, Lehman SM. Colour and odour drive fruit selection and seed dispersal by mouse lemurs. Scientific Reports. 2013;**3**:2424. DOI: 10.1038/srep02424

[34] Marsh KJ, Wallis IR, Kulheim C, Clark R, Nicolle D, Foley WJ, et al. New approaches to tannin analysis of leaves can be used to explain in vitro biological activities associated with herbivore defence. New Phytologist. 2020;**225**:488-498. DOI: 10.1111/nph.16117

[35] Vongsak B, Sithisarn P, Mangmool S, Thongpraditchote S, Wongkrajang Y, et al. Maximizing total phenolics, total flavonoids contents and antioxidant activity of Moringa oleifera leaf extract by the appropriate extraction method. Industrial Crops and Products. 2013;**44**:566-571. DOI: 10.1016/j.indcrop.2012.09.021

[36] Mediani FA, Khatib A, Tan CP. Cosmos Caudatus as a potential source of polyphenolic compounds: Optimisation of oven drying conditions and characterisation of its functional properties. Molecules. 2013;**18**:10452-10464. DOI: 10.3390/molecules180910452

[37] Abdullah S, Shaari AR, Azimi A. Effect of drying methods on metabolites composition of Misai Kucing (Orthosiphon stamineus) leaves. APCBEE Procedia. 2012;**2**:178-182. DOI: 10.1016/j.apcbee.2012.06.032

[38] Borhan MZ, Ahmad R, Rusop M, Abdullah S. Impact of Nano powders on extraction yield of Centella asiatica. Advances in Materials Research. 2013;**667**:246-250. DOI: 10.4028/www.scientific.net/AMR.667.246

[39] Martel J, Ojcius DM, Ko YF, Young JD. Phytochemicals as prebiotics and biological stress inducers. Trends in Biochemical Sciences. 2020;**45**:462-471. DOI: 10.1016/j.tibs.2020.02.008

[40] AA AL-S, Khalil S, EOS H, Attia YA. Effects of fennel seed powder supplementation on growth performance, carcass characteristics, meat quality, and economic efficiency of broilers under thermoneutral and chronic heat stress conditions. Animals. 2020;**10**:206. DOI: 10.3390/ani10020206

[41] Aljumaah MR, Suliman GM, Abdullatif AA, Abudabos AM. Effects of phytobiotic feed additives on growth traits, blood biochemistry, and meat characteristics of broiler chickens exposed to salmonella typhimurium. Poultry Science. 2020;**99**:5744-5751. DOI: 10.1016/j.psj.2020.07.033

[42] Khan UA, Rahman H, Niaz Z, Qasim M, Khan J. Antibacterial activity of some medicinal plants against selected human pathogenic bacteria. European Journal of Microbiology & Immunology. 2013;**3**:272-274. DOI: 10.1556/EuJMI.3.2013.4.6

[43] Wenk C. Herbs and botanicals as feed additive in monogastric animals. Asian-Australasian Journal of Animal Science. 2003;**16**:282-289. DOI: 10.5713/ajas.2003.282

[44] Naiel MA, Ismael NE, Negm SS, Ayyat MS, Al-Sagheer AA. Rosemary

leaf powder-supplemented diet enhances performance, antioxidant properties, immune status, and resistance against bacterial diseases in Nile tilapia (Oreochromis niloticus). Aquaculture. 2020;**526**:735370. DOI: 10.1016/j.aquaculture.2020.735370

[45] Abdul Kari Z, Wee W, Mohamad Sukri SA, Che Harun H, Hanif Reduan MF, Irwan Khoo M, et al. Role of phytobiotics in relieving the impacts of Aeromonas hydrophila infection on aquatic animals: A mini-review. Frontiers in Veterinary Science. 2022;**9**:1023784. DOI: 10.3389/fvets.2022.1023784

[46] Halliwell B. Antioxidant characterization. Methodology and mechanism. Biochemical Pharmacology. 1995;**49**(10):1341-1348

[47] Nile SH, Keum YS, Nile AS, Jalde SS, Patel RV. Antioxidant, anti-inflammatory, and enzyme inhibitory activity of natural plant flavonoids and their synthesized derivatives. Journal of Biochemical and Molecular Toxicology. 2018;**32**:1. DOI: 10.1002/jbt.22002

[48] McCutcheon SC, Jorgensen SE. Phytoremediation. In: Jorgensen SE, Fath BD, editors. Encyclopedia of Ecology. 1st ed. Oxford: Academic Press; 2008. pp. 2751-2766 Available from: http://www.sciencedirect.com/science/article/pii/B9780080454054000690

[49] Zeb A. Concept, mechanism, and applications of phenolic antioxidants in foods. Journal of Food Biochemistry. 2020;**00**:e13394. DOI: 10.1111/jfbc.13394100

[50] Tsao R, Deng Z. Separation procedures for naturally occurring antioxidant phytochemicals. Journal of Chromatography. B, Analytical Technologies in the Biomedical and Life Sciences. 2004;**812**:85-99. DOI: 10.1016/j.jchromb.2004.09.028

[51] Perron NR, Brumaghim JL. A review of the antioxidant mechanisms of polyphenol compounds related to iron binding. Cell Biochemistry and Biophysics. 2009;**53**(2):75-80

[52] Liu ZY, Wang XL, Ou SQ, Hou DX, He JH. Sanguinarine modulate gut microbiome and intestinal morphology to enhance growth performance in broilers. PLoS One. 2020;**15**:e0234920. DOI: 10.1371/journal.pone.0234920

[53] Kikusato M. Phytobiotics to improve health and production of broiler chickens: Functions beyond the antioxidant activity. Invited Review. Animal Bioscience. 2021;**34**(3):345-353. DOI: 10.5713/ab.20.0842

[54] Lillehoj H, Liu Y, Calsamiglia S, et al. Phytochemicals as antibiotic alternatives to promote growth and enhance host health. Veterinary Research. 2018;**49**:76. DOI: 10.1186/s13567-018-0562-6

[55] Valenzuela-Grijalva NV, Pinelli-Saavedra A, Muhlia-Almazan A, Domínguez-Díaz D, González-Ríos H. Dietaryinclusioneffectsofphytochemicals as growth promoters in animal production. Journal of Animal Science and Technology. 2017;**59**:8. DOI: 10.1186/s40781-017-0133-9

[56] Li L, Sun X, Zhao D, Dai H. Pharmacological applications and action mechanisms of phytochemicals as alternatives to antibiotics in pig production. Frontiers in Immunology. 2021;**12**:798553. DOI: 10.3389/fimmu.2021.798553

[57] Chen G, Shui S, Chai M, Wang D, Su Y, Wu H, et al. Effects of paper mulberry (Broussonetia Papyrifera) leaf extract on growth performance and Fecal microflora

of weaned piglets. BioMed Research International. 2020;**2020**:6508494. DOI: 10.1155/2020/6508494

[58] Ndomou SCH, Tiwo TC, Djikeng TF, Teboukeu BG, Kohole HAF, Ngoudjou TD, et al. Correlation between performance, carcass traits and blood lipid profile of broilers chicken fed with rations supplemented with plant powders. Journal of Applied Life Sciences International. 2018;**17**(2):1-12. DOI: 10.9734/JALSI/2018/40681

[59] Lee SH, Lillehoj HS, Hong YH, Jang SI, Lillehoj EP, Ionescu C, et al. In vitro effects of plant and mushroom extracts on immunological function of chicken lymphocytes and macrophages. British Poultry Science. 2010;**51**:213-222

[60] Lee SH, Lillehoj HS, Jang SI, Lee KW, Park MS, Bravo D, et al. Cinnamaldehyde enhances in vitro parameters of immunity and reduces in vivo infection against avian coccidiosis. The British Journal of Nutrition. 2011;**106**:862-869

[61] Shibata T, Nakashima F, Honda K, et al. Toll-like receptors as a target of food-derived anti-inflammatory compounds. The Journal of Biological Chemistry. 2014;**289**:32757-32772. DOI: 10.1074/jbc.M114.585901

[62] Kim DK, Lillehoj HS, Lee SH, Lillehoj EP, Bravo D. Improved resistance to Eimeria acervulina infection in chickens due to dietary supplementation with garlic metabolites. The British Journal of Nutrition. 2013;**109**:76-88

[63] Kim DK, Lillehoj HS, Lee SH, Jang SI, Bravo D. High-throughput gene expression analysis of intestinal intraepithelial lymphocytes after oral feeding of carvacrol, cinnamaldehyde, or capsicum oleoresin. Poultry Science. 2010;**89**:68-81

[64] Misiak B, Łoniewski I, Marlicz W, et al. The HPA axis dysregulation in severe mental illness: Can we shift the blame to gut microbiota? Progress in Neuro-Psychopharmacology & Biological Psychiatry. 2020;**102**:109951. DOI: 10.1016/j.pnpbp.2020.109951

Chapter 3

The Role of *Drosophila melanogaster* (Fruit Fly) in Managing Neurodegenerative Disease in Functional Food and Neutraceuticals Research

Abiola M. Ayodele-Asowata, Ezekiel Olumoye Oyetunji and Babawale Peter Olatunji

Abstract

Fruit fly (*Drosophila melanogaster*) has emerged as a very useful model of neurodegenerative disease and could be more effective for therapeutic screening for neuroprotective properties of functional food and nutraceuticals. There have been no adequate screening models on functional food research in Africa. Limited studies have been reported on the use of *D. melanogaster* an alternative to the use of rodents and other animals in therapeutic screening of functional foods and nutraceuticals. The genomic similarities between *D. melanogaster* and humans, quick generation time, low maintenance requirements, and the accessibility of effective genetic tools, make the fruit fly a suitable research subject for complicated neurodegenerative ailments. However, there is more to be done in understanding complexity in human disease modeling, where the use of fly models will be the best alternative has not been explored. More outcry to conduct studies in disease-related models, the chronic diseases, such as cancer, GI disorders, and cardiovascular diseases, which are causes of death in most industrialized countries are required, although most of the diseases are linked with the intake of dietary fruits, vegetables, and whole grains. So the role of research models cannot be overemphasized, more studies are expected in finding better alternatives to the use of animals in the study of neurodegenerative diseases.

Keywords: drosophila, neurodegenerative, disease, nutraceuticals, functional foods

1. Introduction

In the prevention of chronic diseases such as neurodegenerative diseases, nutrition is critical; not just to meet nutritional requirements but more importantly to contribute to the total wellness of the consumer by either preventing and/or managing such disease conditions [1]. This has further promoted the concept of functional foods and nutraceuticals. The quality of nutritional studies largely depends on the research question addressed, the experimental design, the statistical power, and the composition of the experimental diets. However, while several studies abound on the huge abundance and diversity of functional foods, especially in tropical parts of the world, there is still a serious limitation to rapid and high experimental screening for neuroprotective properties of several functional foods, especially in developing nations. These limitations include modern, effective and accessible experimental models for rapid screening, cost of research, and ethical issues with animal use among others. The vast majority of nutritional studies in model organism have been conducted in laboratory rodents, such as mice and rats. Nutrient requirements for rodents are relatively well established including energy, lipids, fatty acids, carbohydrates, proteins, and amino acids as well as vitamins, minerals, and trace elements, but few cases are reported for the fruit flies [2, 3].

Fruit fly (*D. melanogaster*) has emerged as a very useful model of neurodegenerative disease and could be more effective for therapeutic screening for neuroprotective properties of functional food and nutraceuticals, especially from developing countries of tropical Africa [4]. *D. melanogaster* enables the possibility to conduct studies in disease-related models, not all experiments are required to be conducted with animal models, and therefore this review will focus on compiling scientific works done in the therapeutic screening for neuroprotective properties of functional food and nutraceuticals using *D. melanogaster*.

2. Functional foods and nutraceuticals

Food can be considered functional when it has a significant health effect that extends beyond basic traditional nutrition. Nutraceutical products are derived from foods containing the essential components, such as functional foods, which have therapeutic effects. Its beneficial components can be isolated and purified from plant, animal, or marine sources. Functional foods and nutraceuticals have attracted international interest and shaped a growing global market. They are often known as "functional foods;" nutraceuticals have led to intense controversy due to the fact that they blur the conventional dividing line between diet and medicine. Therefore, functional food provides the human body with the required quantity of basic needs and essential for healthy survival such as proteins, fats, carbohydrates, and vitamins [1]. It is considered "nutraceutical" when functional food helps in disease/disorder prevention and/or medication other than deficiency conditions such as anemia and other neurodegenerative diseases. Therefore, functional food may be used as a nutraceutical to another consumer [5].

When describing nutraceutical, they include products that come naturally with driven nutrients, dietary supplements, herbal products, and generally processed foods. While food nutraceuticals are generally dietary supplementary that are needed for human development. Dietary supplements give direction in which human develops; some epidemiological studies have emphasized on the relationship that exists between disease, lifestyle, and diet [6].

Hippocrates said, "Let your food be your medicine and your medicine be your food." Nutraceuticals play an important role in biological processes such as cell proliferation, antioxidant defense, and gene expression. Nutraceuticals can delay the aging process and decrease the risk of situations such as cancer, heart disease, hypertension, excessive weight, high cholesterol, diabetes, osteoporosis, arthritis, insomnia, cataracts, constipation, indigestion, and many other lifestyle-related disorders [7].

Nutraceuticals are often said to be products that are extracted or purified from animal, plant, or marine sources, which have shown physiological benefits or are known to protect against chronic diseases.

Many studies have been published on therapeutic properties of functional foods, especially those of tropical African origins. In the last few decades, many interesting research publications have originated from Africa on therapeutic properties of several tropical functional foods. However, one major limitation to full evolution of functional food research in Africa has been adequate screening models [7].

Many studies have been carried out based on *in vitro* and *vivo* studies using of mouse and rats, but limited studies have been reported on the use of *D. melanogaster*, while we all know that not all *vivo* studies requires the use of rats, mouse, fish, and others.

Therefore, it is very important to explore more alternative studies tools. Fruit fly has emerged as the new alternative tool in the world for therapeutic screening of functional foods and nutraceuticals.

3. Classification of Functional foods and nutraceuticals

Depending on different parameters, many classifications of nutraceuticals were proposed. Depending on the established stage of nutraceuticals, nutraceuticals are classified as the following:

1. Nutraceuticals consist of compounds that offer therapeutic benefits in many types of research/epidemiology studies but are lacking compliance with large-scale clinical research.

2. Established nutraceuticals consist of compounds that exhibit health benefits well supported by clinical data [2].

They can be divided based on the nutraceutical source from which they are extracted or isolated:

1. Phytochemicals: extracted from plants or herbs, such as flavonoids.

2. Microbial extracted nutrients: such as vitamin A.

3. Nutrients of animal origin: extracted from livestock.

Nutraceuticals, generally, consist of polyunsaturated fatty acids, prebiotics, flavonoids, and vitamins.

Additional nutraceuticals are classifiable as the following:

- *Nutritional enhancements*: These formulas contain nutrients, for example, salt alone or with various preservatives.

- *Functional foods*: These are defined as food enhanced by promoters who promote optimum health and assist in decreasing the disease risk, such as oatmeal containing soluble fiber that reduces cholesterol levels. They are not just nutrients; in brief, they are foods enhanced with nutrients for health benefits [6].

Different researches have emerged that foods can be directly linked to neuro studies and balance of anxiety in the body, for example, potatoes, soybeans lentils, navy beans, yeasts, beans, chickpeas, kidney beans, catfish, milk, eggs, and beef are good sources of lysine. Studies have shown that lysine and amino acid inhibits the hyperactivation of serotonin receptors, which helps to induce anxiety-induced disorders that are caused by serotonin receptors in the intestinal tract due to their hyperstimulation [6].

Valine in its volume and shape roughly looks like threonine and it exhibits some stimulatory effects. Valine has important functions such as growth and repair of tissues, muscle metabolism, and nitrogen balance in the body. Valine is rich in some foods, such as mushrooms, sesame seeds, soy, peanuts, lentils, meat, fish, and cheese. Alanine removes toxins released during muscle breakdown so that the liver can metabolize them and remove them from the body. The cholesterol levels in the body are also checked by alanine [8].

In the human body, the main functions of aspargine are to support the nervous system to maintain equilibrium and regulate metabolism, and they also act as detoxifiers. Aspartic acid has its effect on cellular energy, so it is used to fight fatigue and depression. Sometimes aspartic acid plays a crucial role in manufacturing of other amino acids [9].

Cysteine has multiple functions, such as scavenging free radicals (as an antioxidant) and detoxifiers in the body. It is essential for healthy hair, nails, and skin.

The chronic diseases, such as cancer, GI disorders, and cardiovascular diseases, which are causes of death in most industrialized countries, and these high-risk conditions are reduced significantly with intake of dietary fruits, vegetables, and whole grains (**Table 1**).

3.1 Nutraceutical future

Nutraceutical is frequently referred to in the 21st century as a more attractive functional food. By using nutraceutical tools, the physician of the future would have been a better source to offer preventive medical approaches. Nutraceuticals' advances will encourage individualized nutrition personalized to the profile of a person to maximize health and comfort. The nutraceutical market shows that consumers are looking for minimal foods with additional dietary benefits and organoleptic value. In turn, this progress propels the expansion of global nutraceutical markets. In the new millennium, the evolving nutraceutical manufacturing appears destined to occupy the landscape. Its enormous growth and evolution have consequences for food, healthcare, industries of agriculture, and pharmaceutical [2].

4. *D. melanogaster* as a therapeutic screening model for functional food

4.1 Life cycle of *D. melanogaster*

From egg fertilization to adulthood, the drosophila life cycle spans roughly 10 – 12 days at 25°C. It is a holometabolous insect with significant physical variations between the larvae and adult (metamorphosis) (**Figure 1**). *D. melanogaster* is a model organism utilized mainly in developmental biology [14]. Four developmental stages

	Phytochemicals in functional foods	Functional foods and neutraceuticals	Management of neurodegenerative diseases	References
1	Valine and alanine	Mushrooms, sesame seeds, soy, peanuts, lentils, meat, fish, and cheese.	Valine has important functions such as growth and repair of tissues, muscle metabolism, and nitrogen balance in the body.	[8]
			Alanine removes toxins released during muscle breakdown so that the liver can metabolize them and remove them from the body. The cholesterol levels in the body are also checked by alanine.	[8]
2	Cystine and glutamine		Boosting the immune system, good digestion, and proper brain function, suppression of hunger.	[10]
3	Proline	A meal with high carbohydrate levels.	For proper function of the brain and for formation of chemicals that enhance mental stability and mood serine is very essential.	[11]
4	Tyrosine and tryptophan	Sea vegetables, bananas, peanuts, cabbage mushrooms, watercress, parsley, and seeds.	Can promote good brain chemistry and influence the mood.	[12]
5	Selenocysteine	Rice, nuts, corn, oats, wheat, Brazil nuts, soybeans, chicken, cheese, fish, and seafood.		
6	Vitamin C, Rutin, Ursolic acid, Lutein	Citrus fruits	which are very powerful and reduce the risk of heart stroke, coronary disease, oxidative stress, and inflammation. Citrus fruits help blood vessels to flow continuously so that sufficient oxygen is supplied to brain. Neuroprotection against cognitive deficits and brain damage.	
7	Polyphenols	Apples	Some of the inflammatory damage to brain function can be reduced by flavonoids present in apples. Apples have antioxidants and quercetin, a photochemical that is good for the skin.	
8	Resveratrol and Apolyphenol	Grape juice	which is present in grape juice, has many functions, such as the suppression of degenerative nerve disease, Alzheimer's disease, viral infections, and cancer. Also help to coordinate cognitive functions.	[13]

	Phytochemicals in functional foods	Functional foods and neutraceuticals	Management of neurodegenerative diseases	References
9	Gallic acid	Straw berries	Beneficial effect on behavioral impairments after brain injury induced by ischemia/ reperfusion; cerebroprotective properties.	

Table 1.
Phytochemicals in functional foods and nutraceuticals linked to control neurodegenerative diseases.

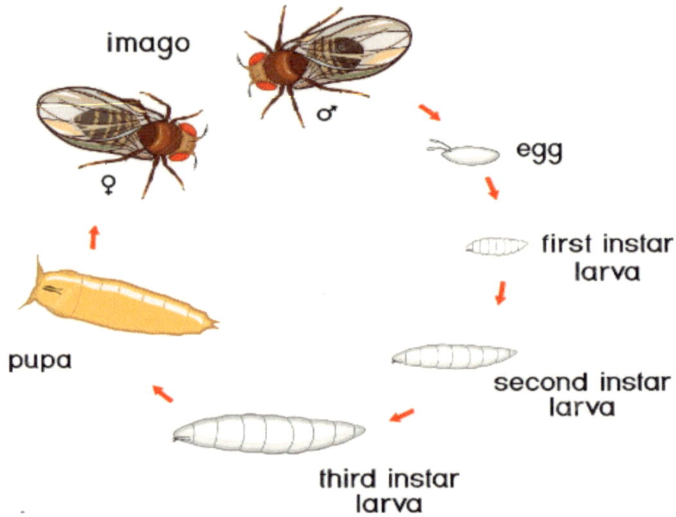

Figure 1.
Life cycle of Drosophila melanogaster.

are included throughout the *Drosophila* life cycle. Embryogenesis is a quick process that is finished 24 hours after the male sperm inseminates the oocyte. A one-cell embryo quickly develops into a syncytial embryo. Nuclear divisions and fast DNA replication take place in the early embryonic syncytium, producing up to 5000 nuclei per embryo. After nuclei migrate to the syncytial blastoderm's periphery and undergo a process known as cleavage to produce the blastoderm, cellularization takes place. Three substages of the larval stage—totaling three instars—take place during the course of around 4 days (**Figure 1**). The majority of cell types are already functionally differentiated and growing as larvae. As a result, many biological queries can already be answered at the larval stage. For example, larvae have proved crucial for research on neural processes, such as memory formation. A simpler example is provided by the larval fly's central nervous system, which has only 10,000 neurons compared to the adult fly's more than 250,000 [15]. After the 3rd instar larva is encapsulated, the pupal stage begins and lasts for around fourth days. New structures are created after the lysis of many larval structures.

The imaginal discs, which were created from undifferentiated larval cells, generate new structures. The adult head, legs, wings, thorax, and reproductive organs are derived from the imaginal discs. After the pupal case has closed, the adult fly comes out. The life expectancy is approximately 30 days, though it can vary depending on the climate. *D. melanogaster* is a good genetic model with relatively easy development in lab settings due to the huge number of eggs laid per female (100 eggs per day), which results in a large offspring after genetic crosses. Important advancements in the understanding of basic biological processes including aging, circadian rhythms, and behavioral research have resulted from the study of adult flies.

4.2 Reports on potentials of *D. melanogaster* as model organism

D. melanogaster, a fruit fly, is used as a model organism to research a variety of topics, from basic genetics through the development of tissues and organs. The human genome is 60% similar to the Drosophila genome, which is less redundant and shares 75% of the genes with flies that cause human disorders [16]. These characteristics, along with a quick generation time, low maintenance requirements, and the accessibility of effective genetic tools, make the fruit fly a suitable research subject for complicated processes important to biological and biomedical studies.

Model for insect control: For almost a century, scientists have utilized the common fruit fly, *Drosophila melanogaster*, as a model organism [17]. Thomas Hunt Morgan (1866–1945) and his students' early research on the *D. melanogaster* resulted in important findings including sex-linked inheritance and the genetic mutations caused by ionizing radiation [18]. The first significant complex organism with its genome sequenced was *D. melanogaster* [19]. Most mammalian genes with *D. melanogaster* orthologs have been discovered to be crucial for typical mammalian development and function. The genomic similarities between *D. melanogaster* and humans have made *D. melanogaster* a more useful model for studying human biology and disease mechanisms. The majority of the genetic material is carried by three of the four chromosomes in the *D. melanogaster* genome, which contains more than 14,000 genes [20, 21]. According to estimates by Lloyd and Taylor [22], the fly possesses functional orthologs of close to 75% of disease-related genes in humans. At the nucleotide or protein sequence level, the overall similarity between fly and mammal homologs is typically around 40%; but, in conserved functional domains, it can be as high as 80–90% [23]. *D. melanogaster* has been used for purposes more than only genetic analysis.

It has been shown to be helpful for pharmacological research. Numerous drug-induced effects that were initially discovered in *D. melanogaster* have been confirmed in mammals [24, 25].

Drosophila melanogaster is a tractable model system for human disease and is used to analyze host interactions with recognized insect infections. Important information on the pathology of infection is provided by studies using insect pathogens6. In the food and nutrition industries, *Drosophila melanogaster* has been used as a model system to assess the possible health advantages of organic foods [26]. *D. melanogaster*-derived recombinant acetylcholinesterase (DmAChE) can be used to identify organophosphate and carbamate pesticides in food, vegetables, and the environment [27]. The creation of a three-electrode biosensor, as a new disposable screen-printed electrode, for quick detection of organophosphate and carbamate pesticides in vegetable and water samples, combines recombinant *D. melanogaster* AChE (R-DmAChE), multi-walled carbon nanotubes, and Prussian blue [28]. Alkylating compounds and other chemicals were tested on the *D. melanogaster* to see if they had any negative effects after numerous generations [17]. A Drosophila model was used to examine the toxicity and mutagenicity of the chemicals 1, 2,4,5-tetrachlorobenzene, 1,4,5-trichloro-2,6-nitrobenzene, pentachloronitrobenzene, methyl-l-(butyl-carbamoyl)-2-benzimidazole carbamate fungicides, and dimethyl-2,3,5,6-tetrachloroterephthalate [17]. Additionally, *D. melanogaster* has been used in resistance research on the insecticides cyclodiene and phenylpyrazole [29]. Endosulfan has negative effects on *D. melanogaster* both at the cellular and organismal levels. This research has shown that the use of *D. melanogaster* as animal model is a fantastic alternative for assessing the risk posed by environmental contaminants [30].

Alternative model organism in nutrigenomics: Currently, nutrigenomics encompasses not only nutrient-gene interactions but also nutrient-epigenetic, nutrient-proteomic, and nutrient-metabolomic interactions, as well as host-diet-microbiome interactions [31, 32]. This is how nutrigenomic research sits at the nexus of diet, health, and genomes [33, 34]. The model is the *Drosophila melanogaster*. Normal nucleotide and protein sequence homology between fly and mammalian species is around 40%; in some conserved functional areas, it can be as high as 90% [23, 35]. For the creation of mutant *Drosophila melanogaster*, chromosomal deletions and mutations targeting more than 80% of its genome have been created [36]. The advantages of *Drosophila melanogaster*, in addition to its well-characterized genome and the good availability of mutant and transgenic flies, include a quick life cycle (12 days for the succession of egg, maggot, pupa, and imago), a short lifespan (roughly 70–80 days), a small size (the ability to breed hundreds of individuals in small bottles), and a relatively simple generation of mutant animals in comparison to other organisms. Particularly, because of the presence of a fat body with adipocytes and conserved metabolic pathways involved in fat metabolism and insulin signaling; particularly, because of the presence of a fat body with adipocytes and conserved metabolic pathways involved in fat metabolism and insulin signaling.

Numerous studies using *Drosophila melanogaster* have examined problems connected to fat, such as cancer or cardiovascular failure [37, 38]. Genetic polymorphisms in the insulin/insulin-like growth factor signaling (IIS), and target of rapamycin (TOR) signaling pathway genes have been linked to changes in triglyceride levels and lipid storage brought on by consumption of high-fat and high-sugar diets [39, 40]. Due to its architecture and similar functions to those of mammals, the fruit fly also resembles a useful model for studying various tissues or organs. The Vienna Drosophila Research Center created an accessible RNAi knockdown fly line collection that targets around 90% of the fly genome, which is a significant accomplishment in *D. melanogaster* genetics research [23]. Caenorhabditis elegans has been used primarily for large-scale RNAi screens of gene function, despite the fact that this organism exhibits systemic RNAi, which makes it impossible to associate the gene interference with a particular cell type [41]. RNAi in *Drosophila melanogaster* can be activated by introducing a transgenic long double-stranded "hairpin" RNA because it is cell autonomous [41]. This tool can be used in conjunction with the GAL4/UAS system in *Drosophila* to inhibit the expression of a particular gene in a variety of cell types, making it possible to create conditional transgenic fly models [42]. This facilitates the study of the over- or under-expression of fly homolog genes and proteins, assisting in the development of fly models for the investigation of human diseases [43].

Model for toxicological studies: In the fields of genetics, biochemistry, cell biology, and developmental biology, *D. melanogaster* is frequently employed as a model organism. It has been employed as a model to clarify human diseases in recent decades, as well as incipiently for toxicological studies [44–46].

This fly has traditionally served as a genotoxicity model, but only recently has it been considered as a possible model for researching systemic toxicology or as a substitute model for toxicology research [44, 45]. The anatomical characteristics of *D. melanogaster* include wings and complex eyes. *D. melanogaster* is an important model to understand not only how the genes induce diseases, but also the discovery of the relation of such genes to diseases [20, 23, 47]. Of particular importance, more than 65–70% of human disease genes are present in *D. melanogaster* [23, 20]; Poddig [48, 49]. Due to its tiny body size and brief lifetime, *D. melanogaster* offers quick production times, convenience of usage, and straightforward laboratory maintenance in large quantities.

4.3 Role of *D. melanogaster* in biomedical research as model in functional food research

Drosophila for Cancer modeling: Since most signaling pathways controlling cell growth and invasion in mammals have a role with flies, it is possible to modify these pathways to create models that replicate the biology of tumors in an easy-to-use model organism such as Drosophila [50, 51]. Additionally, a quick assessment of the main role of conserved oncogenes, and tumor suppressor genes in an entire animal were made possible by the combination of genetic screens and the availability of potent recombination procedures [52].

Recent research employing Drosophila imaginal discs also looked at the processes involved in the recruitment of immune cells (macrophages) to the tumor mass, as well as the mechanisms governing the growth of epithelial tumors and their interaction with the surrounding stromal cells and TME [53, 54].

Epithelial tumors in Drosophila: The origin of over 90% of human malignancies is the epithelium [55]. The distinct cell architecture of epithelial tissues, which consists of junctions and apical and basolateral membrane domains is essential for the preservation of cell physiological functions, distinguishes them from other tissues. Early cancer symptoms do include a loss of cell adhesion and polarity, as well as an increase in cell mobility. Drosophila larval imaginal discs are the ideal model system to simulate the onset of epithelial cancer progression because they are a monolayer epithelium that is constrained apically by a squamous epithelium (peripodial membrane) and basally to the notum by a layer of myoblasts embedded in extracellular matrix. In fact, these larval organs resemble mammalian epithelia both morphologically and biochemically [56]. Additionally, the fruit fly preserves the main signaling pathways that control growth in humans, making it possible to use this animal model to analyze the characteristics of cancer [57]. The imaginal wing and eye discs have been successfully employed in recent years to explore tumor growth and invasion, the role of cancer genes, and chemical screens [58].

Model for COVID-19 research: Leading model organism *D. melanogaster* has been used to examine the biochemical and biological properties of human viruses and the effects they have on host cells [59–61]. The versatility of the model system and the viability of research using human viruses are some of the powerful features in Drosophila that shall be beneficial to explore biological events in a precise detail that may be difficult to overcome using higher animal models [16, 23, 59–61]. Therefore, it was anticipated that *D. melanogaster* would make an ideal model organism for study on COVID-19. As demonstrated in the cases of influenza A and dengue viruses [62, 63], this model organism may help us identify factors that affect host susceptibility to SARS-CoV-2 infection and determine whether those factors are clinically responsible for human susceptibility to SARS-CoV-2. As an alternative, one may investigate the mechanisms behind host innate immune activation in response to SARS-CoV-2 infection and see if changes to these processes might have an impact on how the infected hosts respond to the virus. *D. melanogaster* would undoubtedly be a helpful ally in COVID-19 research and the fight against COVID-19 if the correct questions were asked.

Model for anti-nephrolithiasis agents in kidney stone: Calcium oxalate nephrolithiasis, the 80% most common stone subtype, has a complicated origin [64, 65]. Predisposing variables include metabolic disorders such as hypercalcuria, hyperoxaluria, or hypocitraturia, as well as environmental elements such as diet. These etiological variables throw off the biochemical equilibrium of urine, which causes crystallization and the growth of stones [65]. Rat, mouse, porcine, and canine models, among others, have been created in the past for the research of nephrolithiasis [65]. Historically, the rat model of nephrolithiasis has been the most often used. In the rat model, lithogenic substances (ethylene glycol, ammonium chloride, or vitamin D3) are intraperitoneally injected to cause calculus development [66]. The application of this paradigm has led to inconsistent stone production and a variety of consequences. Additionally, the overall model utility is reduced due to the nephrotoxicity of the lithogenic agents [67, 68]. *D. melanogaster* is a particularly potent model due to its high reproductive rate, fully mapped, and largely understood genome, as well as the simplicity of its experimental design. Nephrocytes and Malpighian tubules (MTs), the two distinct parts of the DM renal system, exhibit a striking degree of structural and functional similarities to the human nephron. The fact that the stones created in the MT are also present in the feces may make it possible to detect stone creation *in vivo* without causing any harm [69].

5. Limitation of fruit fly in nutrition research

Even such fly models can have high degrees of conservation and validity, allowing for speedy screening and result interpretation, modeling complex human disease may be somewhat difficult because such fly models often express only particular components of the disease. It is also conceivable to see that some treatments that are harmful to flies may not be in people and vice versa due to variations in metabolism, despite what could seem to be a strong correlation between the toxicity of the two creatures [70].

The possibility that crucial pathogenetic elements are vertebrate-specific and might be overlooked in invertebrate models is a clear drawback of utilizing fly models. For instance, *Drosophila melanogaster* cannot provide a compelling model for immunological illnesses such as multiple sclerosis.

6. Composition of experimental diets given to *D. melanogaster*

A diet like cornmeal increases the lifespan of *D. melanogaster*, whereas diets high in free accessible carbs (saccharides) and cholesterol can shorten their lives. Uniform bottles and vials are preferred for simplicity of fly culture and transfer. There are numerous common *D. melanogaster* media compositions. For instance, in the lab, the flies are cared for and raised on a medium of cornmeal. (1%, w/v brewer's yeast; 2%, w/v sucrose; 1%, w/v powdered milk; 1%, w/v agar; 0.08%, v/w nipagin) kept at a constant temperature and humidity (23 + 1°C and 60% relative humidity, respectively) for 12 hours of darkness and light. Li et al. [71] and Peng et al. [72] have used basal diets containing 105 g of cornmeal, 21 g of yeast, 105 g of glucose, and 13 g of agar; then, 0.4% of Ethyl 4-hydroxybenzoate was added to the diet in order to prevent mold growth [73].

Based on the dietary components specified in the study procedures, an investigation by a team of researchers showed the nutritional composition of more than 70 published diets used for *D. melanogaster* microbiome research [74]. It demonstrates that there are differences in diets and no fixed ration because the formulation for the recipe's apparent norm is vague. Thus the dietary formulation for *D. melanogaster* may vary in composition in relation to the customary recipe for the study at hand.

7. Conclusion

D. melanogaster models can have high degrees of conservation and validity, allowing for speedy screening and result interpretation, modeling complex human disease may be somewhat difficult because such fly models often express only particular components of the disease. It is also conceivable to see that some treatments that are harmful to flies may not be in people and vice versa due to variations in metabolism, despite what could seem to be a strong correlation between the toxicity and physiology of the two creatures. The possibility that crucial pathogenetic elements related to neurodegenerative diseases and metabolic disorders are vertebrate-specific and might be overlooked in invertebrate models could be a setback in employing the use of fly models. There is need for more research on areas where the uses of fly models have not been explored like inability of provide a compelling model for immunological illnesses such as multiple sclerosis.

Author details

Abiola M. Ayodele-Asowata*, Ezekiel Olumoye Oyetunji and Babawale Peter Olatunji
Faculty of Science, Department of Biosciences and Biotechnology, University of Medical Sciences, Ondo, Ondo State, Nigeria

*Address all correspondence to: mayodeleasowata@unimed.edu.ng

References

[1] Das L, Bhaumik E, Raychudhuri U, Chakraborty R. Role of nutraceuticals in human health. Journal of Food Science and Technology. 2012;**49**:11

[2] Avrelija Cencic WC. The role of functional foods, nutraceuticals, and food supplements in intestinal health. Nutrients. 2010;**2**:15

[3] Hirata T, Cabrero P, Berkholz DS, Bondeson DP, Ritman EL, Thompson JR, et al. In vivo Drosophila genetic model for calcium oxalate nephrolithiasis. American Journal of Physiology-Renal Physiology. 2012;**303**:F1555-F1562

[4] Chi T, Kim MS, Lang S, Bose N, Kahn A, Flechner L, et al. A drosophila model identifies a critical role for zinc in mineralization for kidney stone disease. PLoS One. 2015;**10**:e0124150

[5] Dai Q, Borenstein AR, Wu Y, Jackson JC, Larson EB. Fruit and vegetable juices and Alzheimer's disease: The kame project. The American Journal of Medicine. 2006;**119**(9):751-759

[6] Singh Meerza A, Razak Supraj R, Viswanath SB, Begum PS, Rajagopal S. The impact of functional food and nutraceuticals in health (chapter 2). In: Handbook of Food Bioengineering. Therapeutic Foods. Meerza Abdul Razak at Rayalaseema Univesity; 2018. pp. 23-47. DOI: 10.1016/B978-0-12-811517-6.00002-7

[7] Oboh G, Ademiluyi AO, Ademosun AO, Ogunsuyi OB, Oladun FL. Fruit fly (Drosophila melanogaster): A viable model for screening tropical functional foods for neuroprotective properties. In: Alzheimer's Disease & Treatment. Open access Ebooks. 2018:14

[8] Zagórski Z, Sehested K. Transients and stable radical from the deamination of α-alanine. Journal of Radioanalytical and Nuclear Chemistry. 1998;**232**:139

[9] D'Aniello A. d-aspartic acid: An endogenous amino acid with an important neuroendocrine role. Brain Research Reviews. 2007;**53**(2):215-234

[10] Couto N, Malys N, Gaskell S, Barber J. Partition and turnover of glutathione reductase from *Saccharomyces cerevisiae*: a proteomic approach. Journal of Proteome Research. 2013;**12**(6):2885-2894

[11] Mothet J-P, Parent A, Wolosker H, Linden DJ, Ferris C, Rogawski M, et al. d-serine is an endogenous ligand for the glycine site of the N-methyl-d-aspartate receptor. Proceedings of the National Academy of Sciences of the United States of America. 2000;**97**:4926

[12] Waider J, Araragi N, Gutknecht L, Lesch K. Tryptophan hydroxylase-2 (TPH2) in disorders of cognitive control and emotion regulation: A perspective. Psychoneuroendocrinology. 2011;**36**:393

[13] Berger M, Gray J, Roth B. The expanded biology of serotonin. Annual Review of Medicine. 2009;**60**:355

[14] Wolpert L, Tickle C, Arias AM. Textbook Principles of Development. 5th ed. Oxford University press; 2015;**2**

[15] Widman N. Odour-taste learning in drosophila larvae. Journal of Insect Physiology. 2017;**106**(Pt 1):47-54

[16] Ugur B, Chen K, Bellen HJ. Drosophila tools and assays for the study of human diseases. Disease Models

& Mechanisms. 2016;**9**:235-244. DOI: 10.1242/dmm.023762

[17] Zolfaghari ER, Syrjanen L, Barker H, Supuran CT, Parkkila S. *Drosophila melanogaster*: a model organism for controlling dipteran vectors and pests. Journal of Enzyme Inhibition and Medicinal Chemistry. 2014;**30**:505-513. DOI: 10.3109/14756366.2014.944178

[18] Stephenson R, Metcalfe NH. Drosophila melanogaster: a fly through its history and current use. Journal of the Royal College of Physicians of Edinburgh. 2013;**43**:70-75

[19] Adams MD, Celniker SE, Holt R, a. The genome sequence of Drosophila melanogaster. Science. 2000;**2000**(287):2185-2195

[20] Reiter LT, Potocki L, Chien S, Gribskov M, Bier E. A systematic analysis of human disease-associated gene sequences in Drosophila melanogaster. Genome Research. 2001;**11**:1114-1125

[21] Weavers H, Prieto-Sánchez S, Grawe F, Garcia-López A, Artero R, Wilsch-Bräuninger M, et al. The insect nephrocyte is a podocyte-like cell with a filtration slit diaphragm. Nature. 2009;**457**:322-326

[22] Lloyd TE, Taylor JP. Flightless flies: Drosophila models of neuromuscular disease. Annals of the New York Academy of Sciences. 2010;**1184**:e1-e20

[23] Pandey UB, Nichols CD. Human disease models in Drosophila melanogaster and the role of the fly in therapeutic drug discovery. Pharmacological Reviews. 2011;**63**(2):411-436. DOI: 10.1124/pr.110.003293

[24] Heberlein U, Tsai LT, Kapfhamer D, Lasek AW. Drosophila, a genetic model system to study cocaine-related behaviors: a review with focus on LIM-only proteins. Neuropharmacology. 2009;**2009**(56):97-106

[25] Kaun KR, Devineni AV, Heberlein U. Drosophila melanogaster as a model to study drug addiction. Human Genetics. 2012;**131**:959-975

[26] Chhabra R, Kolli S, Bauer JH. Organically grown food provides health benefits to Drosophila melanogaster. PLoS One. 2013;**2013**(8):e52988

[27] Li J, Qian B, Yin J. Surface display of recombinant Drosophila melanogaster acetylcholinesterase for detection of organic phosphorus and carbamate pesticides. PLoS One. 2013;**8**:e72986

[28] Tang Z, Chen H, Song S. Disposable screen-printed electrode coupled with recombinant *Drosophila melanogaster* acetylcholinesterase and multiwalled carbon nanotubes for rapid detection of pesticides. Journal of AOAC International. 2011;**94**:307-312

[29] Remnant EJ, Good RT, Schmidt JM. Gene duplication in the major insecticide target site, Rdl, in Drosophila melanogaster. Proceedings of the National Academy of Sciences of the United States of America. 2013;**110**:14705-14710

[30] Sharma A, Mishra M, Shukla AK, et al. Organochlorine pesticide, endosulfan induced cellular and organismal response in Drosophila melanogaster. Journal of Hazardous Materials. 2012;**221-222**:275-287

[31] Nieves B, Wagner AE. Drosophila melanogaster as an alternative model organism in nutrigenomics. Genes & Nutrition. 2019;**14**:14. DOI: 10.1186/s12263-019-0641-y

[32] Kussmann M, Van Bladeren PJ. The extended nutrigenomics - understanding

the interplay between the genomes of food, gut microbes, and human host. Frontiers in Genetics. 2011;**21**:2

[33] Ruden DM, De Luca M, Garfinkel MD, Bynum KL, Lu X. Drosophila nutrigenomics can provide clues to human gene-nutrient interactions. Annual Review of Nutrition. 2005;**2005**(25):499-522

[34] Ruden DM, Lu X. Evolutionary conservation of metabolism explains how drosophila nutrigenomics can help us understand human nutrigenomics. Genes & Nutrition. 2006;**2006**(1):75

[35] Ye J, Cui X, Loraine A, Bynum K, Kim NC, White G, et al. Methods for nutrigenomics and longevity studies in drosophila. Biological Aging: Methods and Protocols. Humana Press; 2007. pp. 111-141. DOI: 10.1007/978-1-59745-361-5_10

[36] Parks AL, Cook KR, Belvin M, Dompe NA, Fawcett R, Huppert K, et al. Systematic generation of high resolution deletion coverage of the Drosophila melanogaster genome. Nature Genetics. 2004;**36**:288-292

[37] Guida MC, Birse RT, Dall Agnese A, Toto PC, Diop SB, Mai A, et al. Intergenerational inheritance of high fat diet-induced cardiac lipotoxicity in drosophila. Nature Communications. 2019;**10**:193

[38] Hemphill W, Rivera O, Talbert M. RNA-sequencing of Drosophila melanogaster head tissue on high-sugar and high-fat diets. G3 (Bethesda, Md). 2017;**8**:279-290

[39] Oldham S. Obesity and nutrient sensing TOR pathway in flies and vertebrates: Functional conservation of genetic mechanisms. Trends in Endocrinology and Metabolism. 2011;**22**:45-52

[40] Smith WW, Thomas J, Liu J, Li T, Moran TH. From fat fruit fly to human obesity. Physiology & Behavior. 2014;**136**:15-21

[41] Dietzl G, Chen D, Schnorrer F, Su KC, Barinova Y, Fellner M, et al. A genome-wide transgenic RNAi library for conditional gene inactivation in drosophila. Nature. 2007;**2007**(448):151-156

[42] Cho KS, Bang SM, Toh A. Chapter 26- lipids and lipid signaling in drosophila models of neurodegenerative diseases. In: Watson RR, De Meester F, editors. Omega-3 Fatty Acids in Brain and Neurological Health: Boston: Academic Press; 2014. p. 327-336

[43] Miller J, Chi T, Kapahi P, Kahn AJ, Kim MS, Hirata T, et al. Drosophila melanogaster as an emerging translational model of human nephrolithiasis. The Journal of Urology. 2013;**190**:1648-1656

[44] Abolaji AO, Kamdem JP, Farombi EO, Rocha JBT. *Drosophila melanogaster* as a promising model organism in toxicological studies. Archives of Basic and Applied Medicine. 2013;**1**:33-38

[45] Paula MT, Zemolin AP, Vargas AP, Golombieski RM, Loreto EL, Flores EMM, et al. Effects of Hg (II) exposure on MAPK phosphorylation and antioxidant system in *D. melanogaster*. Environmental Toxicology. 2013;**29**:621-630. DOI: 10.1002/tox.21788

[46] Sudati JH, Vieira FA, Pavin SS, Dias GRM, Seeger RL, Golombieski R, et al. Valeriana officinalis attenuates the rotenone-induced toxicity in *Drosophila melanogaster*. Neuro Toxicology. 2013;**37**:118-126

[47] Poddighe S, Bhat KM, Setzu MD, Solla P, Angioy AM, Marotta R, et al.

Impaired sense of smell in a drosophila Parkinson's model. PLoS One. 2013;**29**(8):e73156

[48] Fortini ME, Skupski MP, Boguski MS, Hariharan IK. A survey of human disease gene counterparts in the drosophila genome. The Journal of Cell Biology. 2000;**150**:23-30

[49] Fortini ME, Bonini NM. Modeling human neurodegenerative diseases in drosophila: On a wing and a prayer. Trends in Genetics. 2000;**16**:161-167

[50] Millburn GH, Crosby MA, Gramates LS, Tweedie S, Fly Base C. Fly Base portals to human disease research using drosophila models. Disease Models & Mechanisms. 2016;**9**:245-252. DOI: 10.1242/dmm.023317

[51] Zhasmine M, Sollazzo M, Allocca M, Valenza AM, Grifoni D, Bellosta P. Drosophila melanogaster: A model organism to study cancer. Frontier in Genetics. 2019;**51**:1-16. DOI: 10.3389/fgene.2019.00051

[52] Sonoshita M, Cagan RL. Modeling human cancers in drosophila. Current Topics in Developmental Biology. 2017;**121**:287-309. DOI: 10.1016/bs.ctdb.2016.07.008

[53] Herranz H, Eichenlaub T, Cohen SM. Cancer in drosophila: Imaginal discs as a model for epithelial tumor formation. Current Topics in Developmental Biology. 2016;**116**:181-199. DOI: 10.1016/bs.ctdb.2015.11.037

[54] Muzzopappa M, Murcia L, Milan M. Feedback amplification loop drives malignant growth in epithelial tissues. Proceedings of the National Academy of Sciences United States of America. 2017;**114**:E7291-E7300. DOI: 10.1073/pnas.1701791114

[55] Hanahan D, Weinberg RA. The hallmarks of cancer. Cell. 2000;**100**:57-70. DOI: 10.1016/S0092-8674(00)81683-9

[56] Wodarz A, Nathke I. Cell polarity in development and cancer. Nature Cell Biology. 2007;**9**:1016-1024. DOI: 10.1038/ncb433

[57] St. Johnston D. The art and design of genetic screens: Drosophila melanogaster. Nature Reviews. Genetics. 2002;**3**:176-188. DOI: 10.1038/nrg751

[58] Tipping M, Perrimon N. Drosophila as a model for context dependent tumorigenesis. Journal of Cellular Physiology. 2014;**229**:27-33. DOI: 10.1002/jcp.24427

[59] Hughes TT, Allen AL, Bardin JE, Christian MN, Daimon K, Dozier KD, et al. Drosophila as a genetic model for studying pathogenic human viruses. Virology. 2012;**423**(1):1-5. DOI: 10.1016/j.virol.2011.11.016

[60] Firzan N, Rahmatika D, Emran TB, Harapan H. Potential application of Drosophila melanogaster as a model organism in COVID-19-related research. Frontier in Pharmacology. 2020;**11**:1-4. DOI: 10.3389/fphar.2020.588561

[61] Panayidou S, Ioannidou E, Apidianakis Y. Human pathogenic bacteria, fungi, and viruses in drosophila: Disease modeling, lessons, and shortcomings. Virulence. 2014;**5**(2):253-269. DOI: 10.4161/viru.27524

[62] Hao L, Sakurai A, Watanabe T, Sorensen E, Nidom CA, Newton MA. Drosophila RNAi screen identifies host genes important for influenza virus replication. Nature. 2008;**454**(7206):890-893. DOI: 10.1038/nature07151

[63] Sessions OM, Barrows NJ, Souza-Neto JA, Robinson TJ, Hershey CL,

Rodgers MA, et al. Discovery of insect and human dengue virus host factors. Nature. 2009;**458**(7241):1047-1050. DOI: 10.1038/nature07967

[64] Finkielstein VA. Strategies for preventing calcium oxalate stones. Canadian Medical Association Journal. 2006;**174**:1407-1409

[65] Ali SN, Dayarathna TK, Ali AN, Osumah T, Ahmed M, Cooper TT, et al. *Drosophila Melanogaster* as a function-based high-throughput screening model for anti-nephrolithiasis agents in kidney stone patients. Disease Models & Mechanisms. 2018;**11**:1-31

[66] Liu J, Cao Z, Zhang Z, Zhou S, Ye Z. A comparative study on several models of experimental renal calcium oxalate stones formation in rats. Journal of Huazhong University of Science and Technology. Medical Sciences. 2007;**27**:83-87

[67] Khan SR, Glenton PA. Experimental induction of calcium oxalate nephrolithiasis in mice. The Journal of Urology. 2010;**184**:1189-1196

[68] Khan SR, Glenton PA, Byer KJ. Modeling of hyperoxaluric calcium oxalate nephrolithiasis: Experimental induction of hyperoxaluria by hydroxy-L-proline. Kidney International. 2006;**70**:914-923

[69] Cognigni P, Bailey AP, Miguel-Aliaga I. Enteric neurons and systemic signals couple nutritional and reproductive status with intestinal homeostasis. Cell Metabolism. 2011;**13**:92-104

[70] Rand MD. Drosophotoxicology: The growing potential for drosophila in neurotoxicology. Neurotoxicology and Teratology. 2010;**32**:74-83

[71] Li YM, Chan HY, Huang Y, Chen ZY. Green tea catechins upregulate superoxide dismutase and catalase in fruitflies. Molecular Nutrition & Food Research. 2007;**51**:546-554

[72] Peng C, Chan HY, Li YM, Huang Y, Chen ZY. Black tea the aflavins extend the lifespan of fruit flies. Experimental Gerontology. 2009;**44**:773-783

[73] Hugo S, Peter G. Getting started. An overview on raising and handling drosophila. In: Dahmann C, editor. Methods in Molecular Biology: Drosophila: Methods and Protocols. Totowa, NJ: Humana Press Inc.; 2008. pp. 27-43

[74] Lesperance DNA, Broderick NA. Meta-analysis of diets used in *drosophila* microbiome research and introduction of the *drosophila* dietary composition calculator (DDCC). G3 (Bethesda). 2020;**10**(7):2207-2211. DOI: 10.1534/g3.120.401235

Chapter 4

Phytochemicals in Tofu and Its Health Benefits

Hilal Ahmad Punoo, Iqra Qureshi and Asiya Mohammad

Abstract

A diet high in plant-based foods will offer an environment rich in phytochemicals, which are nonnutritive components of plants that have health-protective properties. Genistein and daidzein, two isoflavones, are found in soy as a dietary source. While soy has only recently become popular in the United States and Western Europe, it has been a staple of Southeast Asian diets for about five millennia. Among South east Asian populations, consuming a lot of soy is linked to lower risks of some malignancies and cardiovascular disease. The abundance of phenolic compounds, terpenoids, pigments, and other naturally occurring antioxidants found in nuts, whole grains, fruits, and vegetables has been linked to the prevention and/or treatment of chronic diseases like heart disease, cancer, diabetes, and hypertension as well as other medical conditions. Isoflavones are one of the phytochemicals in soy-based products that may support excellent health and are present in soymilk and tofu.

Keywords: soybean, Tofu, isoflavones, genestein, heath benefits

1. Introduction

Soyabean is one of the most important agricultural products with a high commercial value is soy (*Glycine max*), which has a high nutritional value. It is simple to grow, resistant to insects and pathogens, and convenient for processing and cooking [1]. Isoflavones, a class of plant oestrogens with structural and functional similarities to human oestrogens, are present in adequate quantities in soybeans [2, 3]. Popular soybean product tofu is formed by coagulating soymilk to form a protein matrix. It is a popular meal in the Far Eastern nations of China, Japan, and Korea. It is also being considered as a meat substitute in Western nations including the United States, the United Kingdom, and France [4]. It is a low-calorie diet since it contains much more water and contains important amino acids, fatty acids, calcium, as well as a number of valuable phytochemicals derived from soybeans [5].

For more than two thousand years, people have ingested tofu, which dates back to the Western Han period [6]. Tofu is highly nutritious and packed with soy protein. Bean curd, commonly known as tofu, has been manufactured in China for a very long time [7]. Due to its comparatively high content of proteins, fats, vitamins, minerals, and isoflavones, tofu offers several significant nutritional and physiological advantages [8, 9]. It one of the best plant protein sources, is also a good source of healthy fats, vitamins, and minerals, as well as other bioactive substances including

IntechOpen

isoflavones, soyasaponin, and others [10]. As a result, include tofu in a balanced diet may lower the risk of developing conditions such as cardiovascular disease, high blood pressure, diabetes, and hyperlipidemia [11]. Choosing raw soybean seeds, soaking, grinding, heating soymilk, filtering, adding coagulants, pressing, and packaging are all steps in the multi-step process that produces tofu. **Figure 1** shows a flowchart for the production of tofu and its nutritional advantages. As a result, a

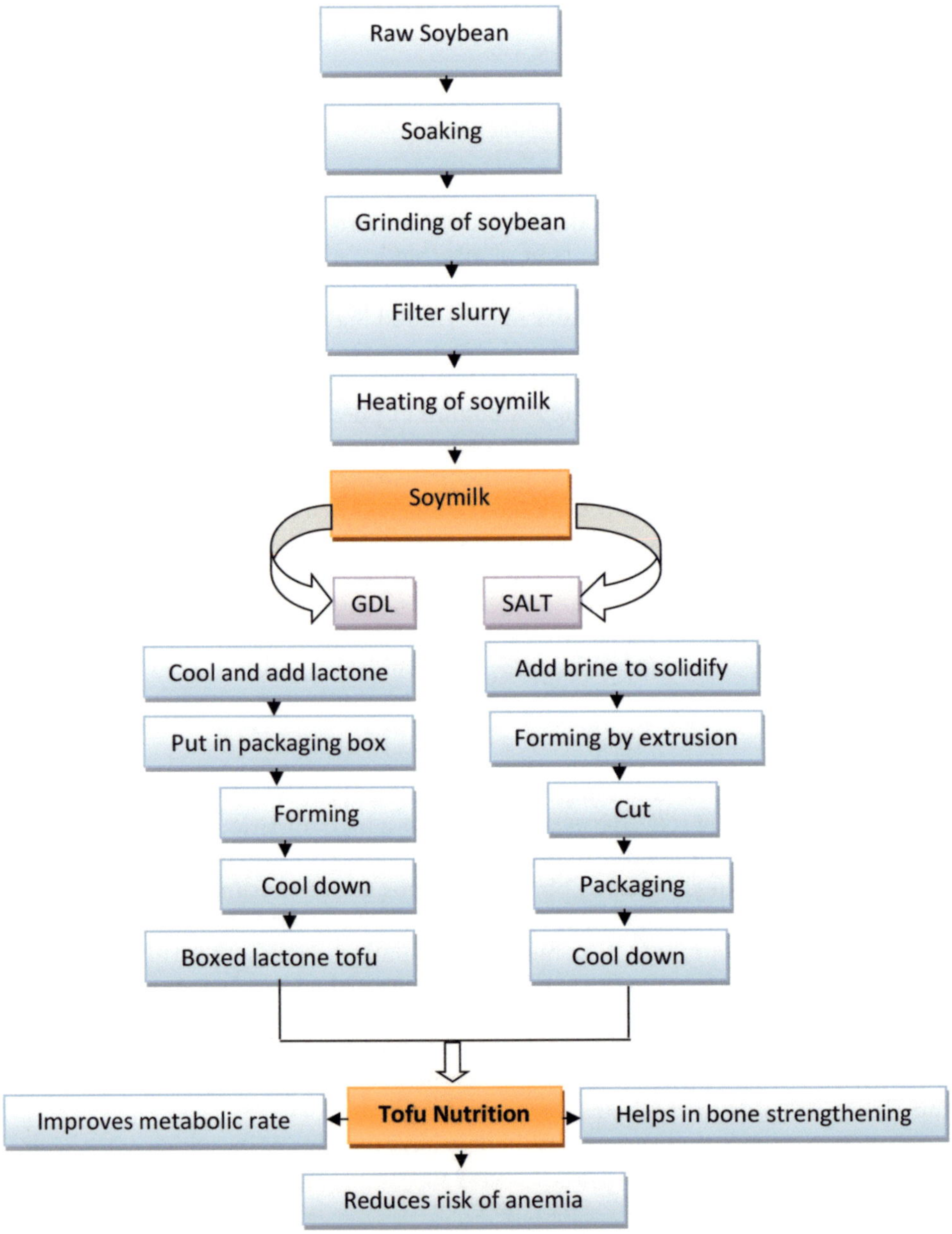

Figure 1.
Flowchart for the production of tofu that includes several steps, and health advantages [12].

variety of factors influence the quality of tofu products. These factors fall into two categories: internal and extrinsic. The variety of soybean seeds, depending on their genotype and protein composition, is an intrinsic factor. Extrinsic elements include food packing and processing environments [13]. Based on product qualities and the various coagulants employed throughout the tofu-making process, tofu products can be divided into firm/soft tofu, packed/pressed tofu, and fermented tofu. The variety of tofu products has fulfilled the various consumption demands, along with the validation of the theory behind tofu formation and the advancement of cutting-edge food processing methods [13–15]. The majority of soy's anti nutritional components may be eliminated through processing, which also greatly increases the soy protein's ability to be digested. According to studies, the digestibility of whole, ripe soybeans is only 65.3%; after being made into tofu and soy milk, it increases to 85% and 92–98%, respectively [16]. The production of tofu is attributed to the gelation properties of soybean protein because tofu is a well-known highly hydrated gel-type food (Singh et al., 2008). The final flavour, quality, and shelf life of the tofu will depend on the materials used for packing and storage product. Products made of tofu can be purchased in bulk, water-filled tubs, plastic bags, or vacuum-sealed packages [17]. Even under refrigeration, the shelf life of tofu products is only a few days because to the high moisture and protein content, which provides an ideal environment for the growth of bacteria [18]. The shelf life of tofu varies significantly in Japan, though, from 1 to 5 days for raw tofu to roughly a week for packed tofu to 3 weeks to a month for pasteurised tofu to 6 months to 2 years for aseptically processed tofu [19].

2. Toffu processing

2.1 Preparation of soymilk

Rekha et al. [20] 200 g of cleaned soybeans were allowed to soak for 12 h at room temperature in excess water. Beans were put directly to boiling water with 1% sodium bicarbonate solution, blanched at 85°C for 5, 10, and 15 min, and then drained. After dehulling each batch by hand, the cotyledons were pulverised in a high-speed blender with water (1.6 L). The soybean slurry was continuously stirred while being indirectly heated in a water bath for 45 min at 85°C. After separating the soymilk from the leftover material (okara) using a double-layered cheesecloth filter, the hot slurry was cooled and the fatty layer was scraped off. The Refractometer's was used to measure the soymilk solid content, which was then corrected with water to 7°, 8°, and 9° Brix.

2.2 Preparation of Tofu

Soymilk (200 ml) was heated to 95°C for 5 min. It is then continuously stirred at ambient temperature while cooling to 80°C. In separate batches, soymilk was mixed with a solution of magnesium chloride (0.2%) and calcium sulphate (0.2%) w/v (1:1) and swirled for 5, 10, 15, and 20 minutes. Without disturbing it, the milk was left to coagulate for 15 min. The coagulated milk was poured into a cheesecloth mould that was lined with porous plastic. The weights used to press the curd were 500, 700, and 1000 g for the first 15 min, followed by 500 g for the following 15 min. After pressing, the cloth was taken off, and the tofu was kept chilled while being preserved in water [20].

2.3 Factors influencing tofu's quality features

The quality of soymilk and the subsequent coagulation process affect the yield and quality of tofu, whereas the quality of soymilk is influenced by the type of soybeans used and the conditions under which it is prepared. The most crucial process in forming tofu is coagulation, which is dependent on several factors, including the soymilk's concentration and temperature, the type and proportions of the coagulant, the mixing technique, etc. The most difficult aspect of manufacturing tofu is determining out the precise amount of coagulant to add to the soymilk because it has a significant impact on the product's yield and quality. The appearance of whey can also be used to estimate the concentrations of coagulants. If the right amount of coagulant is employed, whey turns translucent and has an amber or light yellow tint. However, if too much coagulant is used, the whey will turn yellowish, have a bitter flavour, and have a gritty texture [21].

2.4 Phytochemicals in Tofu

Biologically active, non-nutritive substances from plants are referred to as phytochemicals. The phytochemicals in soybean are very diverse. They consist of trypsin inhibitors, isoflavones, saponins, phytates, phytosterols, and phenolic acids. Diabetes, cancer, and cardio vascular disorders are just a few of the chronic illnesses that phytochemicals may help to prevent [22]. The low prevalence of prostate and breast cancer in Japanese men and women, respectively, is thought to be in part due to soybean consumption. Chinese who frequently eat soybeans and/or tofu have a 50% lower risk of stomach, colon, rectum, breast, and lung cancer compared to Chinese who eat soy or soy products infrequently [23]. The isoflavones genistein and diadzein are found in exceptionally high and unique concentrations in soybeans and non-fermented soy products, such as tofu. In culture, these isoflavonoids prevent the development of cancer cells that are both hormone-dependent and hormone-independent (**Figure 2**) [24].

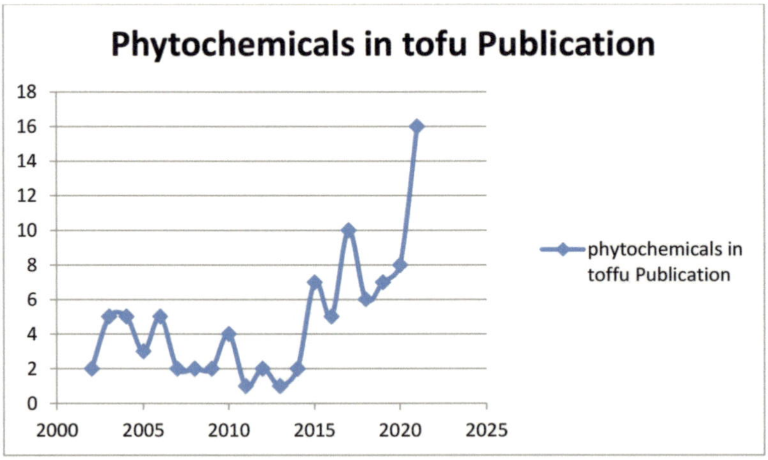

Figure 2.
No. of publications during last 20 years (2002–2022) with the term "Phtochemicals in tofu" (source: Science.gov) use scopus or web of science data.

2.4.1 Isoflavones

Isoflavones which are nearly exclusively found in members of the legume family, are only present in significant quantities in soybeans. The two main isoflavones are genistein and daidzein. Dry soybeans contain an average of 1600–2400 mg/kg of isoflavones. Isoflavones are highly soluble in alcohol, mildly water soluble, and heat stable. Because of rinsing, filtering, or dilution with water or flavourings, tofu and soy milk have decreased isoflavone concentrations [25].

The most prevalent class of phytoestrogens, isoflavones are found in high quantities in soybeans and soy-based foods like tofu. Isoflavones naturally occur in forms called glycosides that are less bioavailable than their aglycone counterparts. By increasing the bioavailability of isoflavones and acting as a bio-catalyst for the conversion of isoflavone glycosides to isoflavone aglycones, β-glycosidase can be utilised to enhance the quality of tofu [26]. Processing tofu is a sophisticated physicochemical process, as is widely known. The distribution of nutritional components and the conditions under which they are processed may have an impact on the tofu's nutritious value. Isoflavones are widely present and explored in beans, grains, and fruits as a form of phytoestrogen with a molecular structure similar to that of oestrogen. Twelve isoflavones, comprising three aglycones (free isoflavones) of genistein, daidzein, and glycitein and their corresponding three glucosidic conjugates, have so far been isolated and characterised in soybean seeds [27]. Because this form of phytoestrogen can reduce the incidence of breast cancer, prostate cancer, cardiovascular disease, osteoporosis, climacteric syndrome and its linked disease, the sources and compositions of isoflavones have been widely explored [28]. The occurrence of intricate physicochemical reactions, the composition of isoflavones in the final product or intermediate product, and the obvious differences between each processing step. By analysing the changes in isoflavone composition brought on by GDL during the preparation of tofu, it was revealed that the processing steps of soaking raw soybean seeds, filtering soybean slurry, and coagulation step had loss ratios of 4%, 31%, and 18%, respectively, of the total mass of isoflavones [29]. After heating soybean slurry to a maximum value for a certain isoflavone, the concentration of aglycones decreased in conjunction with the tofu-making process. Throughout the tofu processing, the amounts of β-glucosides and malonylglucosides were consistently lowered [27]. When soybeans are processed, a huge percent of the isoflavones are lost. This loss occurs in the preparation of tofu (44%), soy isolate (53%), and tempeh (12 and 49% during the soaking and heat processing steps, respectively) [30].

The four chemical forms of soy isoflavones—aglycone (daidzein, genistein, and glycitein), glucoside (daidzin, genistin, and glycitin), acetylglucoside (acetyldaidzin, acetylgenistin, and acetylglycitin), and malonylglucoside—contain a total (malonyldaidzin, malonylgenistin and malonylglycitin) [31]. Various biological properties of soy isoflavones that may include.

2.4.1.1 Soy isoflavones and osteoporosis

Menopause-related ovarian hormone deficit causes an increase in bone turnover and an imbalance between resorption and creation, which speeds up bone loss [32]. In comparison to baseline, soy isoflavones significantly raised bone mineral density in women by 54% and lowered the bone resorption marker urine deoxypyridinoline by 23%. According to a sensitivity study, soy isoflavones had a substantial impact on deoxypyridinoline and bone mineral density. An abrupt decrease in oestrogen levels in ostmenopausal women causes an increase in the pace of bone remodelling, which is linked to a loss of bone mineral density and an increased risk of fractures [33–35].

2.4.1.2 Soy isoflavones' effects on memory and learning

High plasma isoflavone levels from dietary phytoestrogens can have a considerable impact on anxiety, learning, and memory as well as sexually dimorphic brain regions [36]. According to reports, the phytoestrogen genistein, which may cross the blood-brain barrier, has an antioxidant action that protects against the harm caused by ultra-violet (UV) radiation and chemicals. Soy's antioxidant properties may offer protection from neurodegenerative disorders [37].

2.4.1.3 Soy isoflavones' effects on coronary heart disease

Foods containing soy protein are a good source of the phytoestrogen isoflavones genistein and daidzein. Since a high dietary consumption of soy-containing foods has been linked to lower rates of chronic diseases, including coronary heart disease, there is growing interest in these compounds [38].

2.4.1.4 Effect of soy isoflavones on diabetes

Chronic insulin resistance and a loss of functional pancreatic β-cell mass are the causes of type 2 diabetes. Numerous studies conducted over the past ten years have shown that genistein has anti-diabetic properties that are distinct from its roles as an oestrogen receptor agonist, antioxidant, or tyrosine kinase inhibitor, including direct effects on β-cell proliferation, glucose-stimulated insulin secretion, and protection against apoptosis [39].

2.4.2 Genestein

High concentrations of the phytoestrogen genistein, an isoflavone, can be found in soy products. Numerous ailments, including cancer, cardiovascular disease, osteoporosis, and postmenopausal symptoms, are helped by genistein. Genistein has been shown to inhibit cytokine-induced signal transduction processes in immune system cells, which led us to believe that it may also have anti-inflammatory properties. Since as genistein binds to oestrogen receptors and shares structural similarities with oestrogen, it is possible that it exerts an estrogenic effect [40]. Foods from leguminous plants contain genistein in large quantities. The most genistein has reportedly been found in soybeans, a cholesterol-free, high-protein legume. Other legumes like chickpeas (garbanzo beans) contains a small amount of genistein. Soy-based foods, such as tofu, soy milk, soy flour, textured soy protein, soy protein isolates, tempeh, and miso, contain genistein in varying levels [41]. The physiologically active glucoside genistin is the principal dietary source of genistein. The sugar molecule from the isoflavone glycoside, genistin, is released during the fermentation or digestion of soybeans or soy products, leaving the isoflavone aglycone, genistein [42].

2.4.3 Flavnoids

The flavonoids, which constitute up the largest group of plant phenols, are low-molecular-weight chemicals that typically exist linked to sugar molecules. Anthocyanins and anthoxanthins are two types of flavonoids. The molecules of red, blue, and purple pigment known as anthocyanins. Anthoxanthins are colourless or white to yellow compounds that include flavonols, flavones, flavanols, and isoflavones [43].

2.5 Saponins in Tofu

The term saponin has the same origin as soap. When heating raw soymilk to make tofu (bean curd), a lot of bubbles are produced. These bubbles contain significant amounts of saponin components. These bubbles, which have a harsh taste, are skimmed off during regular cooking. In Japan, however, an antifoamer is used throughout the tofu making process rather than their removal. Because saponins are preserved as a result, the traditional tofu production method may have contributed to our good health and longer than average lifespan. Recently, a method without antifoamers that keeps the saponin elements in tofu curd has been developed [44]. Triterpene or steroid glycosides, including steroid alkaloid glycosides, are a more precise definition of saponins. The saponins with pentacyclic triterpenes as the aglycone, especially those of the oleanane type, are the most common [45]. Depending on how many sugar chains are connected to the agylcone, saponins can be further divided into different groups. Saponin concentrations in processed soy foods like tofu and soymilk are only slightly lower than those in soybeans, which contain 1–5% of saponins [46]. Soybeans account for over 80% of all dietary saponins in a normal Japanese diet [47]. Group A saponins have a hydroxyl group at position C-21, whereas group B saponins contain a hydrogen atom there. The carbonyl group at position C-22 distinguishes group E saponins from groups A and B. Group B saponins are referred to as DDMP saponins if they include a 2,3-dihydro-2,5-dihydroxy-6-methyl-4H pyran-4-one (DDMP) moiety at the C-22 position [48].

3. Health effects of soya saponins

Soya saponins aid in antioxidant, hepatoprotective, and anti-cancer properties as well as cardiovascular and cardiovascular protective effects. Their chemical composition heavily influences how they affect human health. Due to this class of soy chemicals' limited absorption, it's possible that the GI tract's indirect activities are what trigger their bioactivity [49].

Cholesterol and some saponins with clearly defined structural properties interact to produce insoluble complexes. This complex-forming activity in the gut prevents both endogenous and exogenous cholesterol from being absorbed via the intestinal tract. (2) By producing mixed micelles, saponins can obstruct the enterohepatic circulation of bile acids. Effectively prevented is the reabsorption of bile acids from the terminal ileum [50].

4. Conclusion

Many people have indicated a strong desire to reduce back on the amount of animal products in their diet for moral, environmental, and health grounds. Tofu is consumed because it is high in nutrients and may be included into vegetarian and hypocaloric diets. Therefore, it is crucial to comprehend the crucial aspects that could affect tofu quality in order to produce high-quality and superbly flavoured tofu goods in order to satisfy the growing client expectations. The variety of soybeans used in the production process, the composition of the soybean protein, its structural characteristics and nutritional value, the type of coagulant used, production technology, and packaging materials are just a few of the variables that affect the final tofu product's quality..Soybeans are unique among the legumes because they are a concentrated source of isoflavones. The evidence supporting the prevention of chronic diseases by soy foods and isoflavones ranges from very well established benefits, such as the decrease of heart disease and reducing cholesterol levels, to highly speculative effects, such as the relief of menopausal symptoms and osteoporosis. Therefore, it is evident that consuming soy foods has more positive impacts than negative effects, even without conclusively proving that doing so lowers the risk of developing chronic diseases.

Author details

Hilal Ahmad Punoo*, Iqra Qureshi and Asiya Mohammad
Department of Food Science and Technology, University of Kashmir, India

*Address all correspondence to: hilal_punoo@rediff.com

References

[1] Wang G, Kuan SS, Francis OJ, Ware GM, Carman AS. A simplified HPLC method for the determination of phytoestrogens in soybean and its processed products. Journal of Agricultural and Food Chemistry. 1990;**38**:185-190

[2] Potter SM, Baum JA, Teng H, Stillman RJ, Shay NF, Erdman JW. Soy protein and isoflavones: Their effects on blood lipids and bone density in postmenopausal women. The American Journal of Clinical Nutrition. 1998;**68**:1375-1379

[3] Adlercreutz H. Phyto-estrogens and cancer. The Lancet Oncology. 2002;**3**:364-373

[4] Sadler MJ. Meat alternatives – Market developments and health benefits. Trends in Food Science and Technology. 2004;**15**:250-260

[5] Watanabe T. Science of Tofu. Kyoto: Food Journal Co. Ltd; 1997. pp. 8-63

[6] Shi YG, Liu LL. Research progress on correlation between soybean protein and tofu quality. Journal of Food Technology. 2018;**36**:1-8

[7] Zhou H, Tian Y, Wang R. Spoilage microorganisms in bean products. In: Wang Y, Zhang WG, Fu LL, editors. Food Spoilage Microorganisms Ecology and Control. Boca Raton, FL: CRC Press; 2017. pp. 431-516

[8] Wang E, Li Y, Maguy BL, Lou Z, Wang H, Zhao W, et al. Separation and enrichment of phenolics improved the antibiofilm and antibacterial activity of the fractions from Citrus medica L. var. sarcodactylis in vitro and in tofu. Food Chemistry. 2019;**294**:533-538

[9] Yang WT, Cho KM, Lee JH. Comparative analysis of isoflavone aglycones using microwave-assisted acid hydrolysis from soybean organs at different growth times and screening for their digestive enzyme inhibition and antioxidant properties. Food Chemistry. 2020;**305**:125462

[10] White LR, Petrovitch H, Ross GW, Masaki K, Hardman J, Nelson J, et al. Brain aging and midlife tofu consumption. Journal of the American College of Nutrition. 2000;**19**:242-255

[11] Takahashi Y, Ishiguro T, Murasawa H, Ishikawa-Takano Y, Watanabe J, Yamaki K. Comparison of the effects of differently processed soy foods on lipid metabolism in rats. Food Science and Technology Research. 2017;**23**(1):163-168

[12] Zheng L, Regenstein JM, Teng F, Li Y. Tofu products: A review of their raw materials, processing conditions, and packaging. Comprehensive Reviews in Food Science and Food Safety. 2020:1541-4337.12640–. DOI: 10.1111/1541-4337.12640

[13] Zhang Q, Wang C, Li B, Li L, Lin D, Chen H, et al. Research progress in tofu processing: From raw materials to processing conditions. Critical Reviews in Food Science and Nutrition. 2018;**58**(9):1448-1467

[14] Dey A, Prasad R, Kaur S, Singh J, Luwang MD. Tofu: Technological and nutritional potential. Indian Food Industry Magazine. 2017;**36**(3):8-24

[15] Shurtleff W, Aoyagi A. History of Tofu and Tofu Products (965 CE to 2013). Lafayette, CA: Soyinfo Center Publishing; 2013

[16] Du LQ. New Technology of Tofu Production. 1st ed. Beijing, China: Chemical Industry Publishing House; 2018. pp. 1-64

[17] Shurtleff W, Aoyagi A. Tofu & Soymilk Production: The Book of Tofu a Craft and Technical Manual. Vol. 2. Lafayette, CA: Soyinfo Center Publishing; 2000

[18] Rossi F, Felis GE, Martinelli A, Calcavecchia B, Torriani S. Microbiological characteristics of fresh tofu produced in small industrial scale and identification of specific spoiling microorganisms (SSO). LWT-Food Science and Technology. 2016;**70**:280-285

[19] Snyder HE, Wilson LA. Processing for the food industry. In: Caballero B, Finglas P, editors. Encyclopedia of Food Sciences and Nutrition. Oxford, UK: Elsevier Science; 2003. pp. 5383-5389

[20] Rekha CR, Vijayalakshmi G. Influence of processing parameters on the quality of soycurd (tofu). Journal of Food Science and Technology. 2013;**50**(1):176. DOI: 10.1007/S13197-011-0245-Z

[21] Prabhakaran M. Isoflavone levels and the effect of processing on the content of isoflavones during the preparation of soymilk and tofu. 2006. Available from: https://core.ac.uk/download/pdf/48628975.pdf

[22] Wang T. Minor constituents and phytochemicals of soybeans. Soybeans. Elsevier; 2008. Available from: https://www.sciencedirect.com/science/article/pii/B9781893997646500135

[23] Messina MJ, Persky V, Setchell KD. Barnes soy intake and cancer risk: A review of the in vitro and in vivo data. Nutrition and Cancer. 1994;**21**(2):113-131

[24] Herman C, Adlercreutz T, et al. Soybean phytoestrogen intake and cancer risk. Journal of Nutrition. 1995;**125**:757S-770S

[25] Anderson RL, Wolf WJ. Compositional changes in trypsin inhibitors, phytic acid, saponins and isoflavones related to soybean processing. The Journal of Nutrition. 1995a;**206**(suppl):575S-586S

[26] Ryu Y-G, Won B, Park H-R, Ghafoor K, Park J. Effects of the β-Glycosidase Reaction on Bio-Conversion of Isoflavones and Quality during Tofu Processing. DOI: 10.1002/jsfa.3894

[27] Kudou S, Fleury Y, Welti D, Magnolato D, Uchida T, Kitamura K, et al. Malonylisoflavone glycosides in soybean seeds (*Glycine max* Merrill). Agricultural and Biological Chemistry. 1991;**55**:2227-2233

[28] Zaheer K, Akhtar MH. An updated review of dietary isoflavones: Nutrition, processing, bioavailability and impacts on human health. Critical Reviews in Food Science and Nutrition. 2015;**2015**. DOI: 10.1080/10408398.2014.989958

[29] Jackson CJC, Dini JP, Lavandier C, Rupasinghe HPV, Faulkner H, Poysa V, et al. Effects of processing on the content and composition of isoflavones during manufacturing of soy beverage and tofu. Process Biochemistry. 2002;**37**:1117-1123

[30] Wang HJ, Murphy PA. Mass balance study of isoflavones during soybean processing. Journal of Agricultural and Food Chemistry. 1996;**44**:2377-2383

[31] Kao TH, Chen BH. Functional components in soybean cake and their effects on antioxidant activity. Journal of Agricultural and Food Chemistry. 2006;**54**:7544-7555

[32] Verhaeghe J, van Bree R, Van Herck E, et al. Effects of recombinant

human growth hormone and insulin-like growth factor-I, with or without 17 beta-estradiol, on bone and mineral homeostasis of aged ovariectomized rats. Journal of Bone and Mineral Research. 1996;**11**:1723-1735

[33] Wei P, Liu M, Chen Y, Chen DC. Systematic review of soy isoflavone supplements on osteoporosis in wome. Asian Pacific Journal of Tropical Medicine. 2012;**5**:243-248

[34] Nielsen TF, Ravn P, Bagger YZ, Warming L, Christiansen C. Pulsed estrogen therapy in prevention of postmenopausal osteoporosis. A 2-year randomized, double blind, placebo-controlled study. Osteoporosis International. 2004;**15**:168-174

[35] Seeman E. Estrogen, androgen, and the pathogenesis of bone fragility in women and men. Current Osteoporosis Reports. 2004;**2**:90-96

[36] Lephart ED, West TW, Weber KS, et al. Neurobehavioral effects of dietary soy phytoestrogens. Neurotoxicological Teratology. 2004;**24**:5-16

[37] Zeng H, Chen Q, Zhao B. Genistein ameliorates beta-amyloid peptide (25-35) -induced hippocampal neuronal apoptosis. Free Radical Biological Medicine. 2004;**36**:180-188

[38] Tikkanen MJ, Adlercreutz H. Dietary soy-derived isoflavone phytoestrogens. Could they have a role in coronary heart disease prevention? Biochemical Pharmacology. 2000;**60**:1-5

[39] Gilbert ER, Liu D. Anti-diabetic functions of soy isoflavone genistein: Mechanisms underlying its effects on pancreatic β-cell function. Food Functions. 2013;**4**:200-212

[40] Verdrengh M, Jonsson IM, Holmdahl R, Tarkowski A. Genistein as an anti-inflammatory agent. Inflammation Research. 2003;**52**(8):341-346. DOI: 10.1007/S00011-003-1182-8/METRICS

[41] Ganai AA, Farooqi H. Bioactivity of genistein: A review of in vitro and in vivo studies. Biomedicine & Pharmacotherapy. 2015;**76**:30-38. DOI: 10.1016/J.BIOPHA.2015.10.026

[42] Jung W, Yu O, Lau S-MC, O'Keefe DP, Odell J, Fader G, et al. Identification and expression of isoflavone synthase, the key enzyme for biosynthesis of isoflavones in legumes. Nature Biotechnology. 2000;**18**:208-212

[43] Hertog MGL, Hollman PCH, Katan MB, Kromhout D. Intake of potentially anticarcinogenic flavonoids and their determinants in adults in the Netherland. Nutrition Cancer. 1993;**20**(1):21-29

[44] Tsukamoto C, Yoshiki Y. Soy Saponin. Soy in Health And Disease Prevention. 2005:155-172. DOI: 10.1201/9781420026566-9

[45] Mahato S, Garai S. In: Herz W, Kirby GW, Moore RE, Steglich W, Tamm C, editors. In Progress in the Organic Chemistry of Natural Products. Vol. 74. New York: Springer; 1998. pp. 1-196

[46] Anderson R, nutrition, WW-T. J. of, & 1995, undefined. Compositional changes in trypsin inhibitors, phytic acid, saponins and isoflavones related to soybean processing; 1995

[47] Messina M, Flickinger B. Hypothesized anticancer effects of soy: Evidence points to isoflavones as the primary anticarcinogens. Pharmaceutical Biology. 2002;**40**(SUPPL.):6-23. DOI: 10.1076/PHBI.40.7.6.9171

[48] Heng L, Vincken J, Hoppe K, Chemistry GVK-F. Stability of Pea DDMP

Saponin and the Mechanism of Its Decomposition. Elsevier; 2006

[49] Kang J, Badger TM, Ronis MJJ, Wu X. Non-Isoflavone phytochemicals in soy and their health effects. Journal of Agricultural and Food Chemistry. 2010;**2010**

[50] Oakenfull D, Sidhu GS. Could saponins be a useful treatment for hypercholesterolaemia? European Journal of Clinical Nutrition. 1990;**44**:79-88

Chapter 5

Food Applications of *Telfairia occidentalis* as a Functional Ingredient and Nanoencapsulation as a Promising Approach toward Enhancing Food Fortification

Aisha Idris Ali, Munir Abba Dandago and Fatima Idris Ali

Abstract

The cucurbitaceous vegetable *Telfairia occidentalis* Hook. f. "Fluted pumpkin" is grown in West Africa, especially in Nigeria for its nutritious leaves and seeds. It has various industrial applications, such as food and medicine. *T. occidentalis* contains essential nutrients that could play a significant role in human nutrition. Based on its chemical composition and nutritional properties, it can be used to overcome malnutrition. *T. occidentalis* leaves and seeds are rich in phenolic compounds, minerals, vitamins, proteins, essential amino acids, and other essential phytochemicals which can play a regulatory and functional role. The leaves and seeds of this plant have also been used in different food applications. This chapter highlights the reported knowledge relevant to the use of *T. occidentalis* as a food fortificant and therapeutic agent based on its prominent biological activities and the presence of phytochemicals. However, conventional food fortification methods do not completely meet the functional requirements for bioactive compounds. They also have unsatisfactory flavor profiles, as well as poor stability and bioavailability. These disadvantages can be mitigated by encapsulating the bioactive components in nanoparticle-based delivery systems. Nanofood fortification has a wide range of advantages in the protection of phytochemicals through the use of an encapsulation technique, and some micronutrients that are rapidly degraded or not properly absorbed by the body can also be aided by food fortification on the nanoscale. Nanosuspensions, nanoemulsions, nanoliposomes, and cyclodextrin carriers are some of the various nanotechnology techniques that can be used for food fortification, which have been discussed in this chapter.

Keywords: fluted pumpkin, minerals, vitamins, *Telfairia occidentalis*, food fortification, nanoencapsulation, nanoliposomes, nanoemulsions, nanosuspensions

1. Introduction

Plants have existed on Earth since ancient times, as depicted in holy books about the creation of the universe. Plants are regarded as immediate companions of humans and are used to meet basic human needs, such as clothing, food, shelter, and medicine. As a result, plants will always be economically, industrially, environmentally friendly, historically, and spiritually important to humanity's survival and advancement. The presence of natural phytochemical compounds in plants is critical for human and animal health [1]. These phytochemical compounds are primarily synthesized in plants *via* primary and secondary metabolic pathways. These are flavonoids, alkaloids, steroids, and glycosides that have a variety of biological activities [2]. Food demand has increased in developing countries in recent decades in order to combat hunger and malnutrition. The Food and Agriculture Organization (FAO) also encourages 70–80% of the world's population to consume medicinal plants in order to reduce the economic cost (25%) of synthesized drugs [3, 4].

Telfairia occidentalis Hook F. (fluted pumpkin) belongs to the *Cucurbitaceae* family. It is a West African native that is primarily grown in Sierra Leone, Ghana, and Nigeria [5]. It is thought to have originated in Nigeria's southeastern region and been spread by the *Igbos*, who have been cultivating this crop since antiquity. Initially, *T. occidentalis* was wild throughout its current range; nevertheless, wild plants may have been harvested to local extinction and are now substituted with cultivated types [6, 7]. *T. occidentalis* is a nutritious vegetable widely cultivated for its palatable and nutritious leaves and seeds and contains essential nutrients that can play a major role in human nutrition [8, 9]. *T. occidentalis* find numerous applications in different industries, such as food and medicine [10–12]. The leaves of *T. occidentalis* are good sources of phenolic compounds, essential amino acids, vitamins, minerals, proteins, antioxidants, and other essential nutrients, which [13] can play a regulatory and functional role. The leaves and seeds of this plant have also been used in different food applications, such as bread, cookies, *chin-chin*, cassava pasta, smoothies, complementary food, and soups condiment (*Ogiri ugu*) [2]. The chemical and nutritional profile of *T. occidentalis* suggests its use to overcome malnutrition.

However, traditional fortification programs have primarily focused on disease eradication through the use of higher concentrations of active ingredients for deficiency disease eradication; however, the focus has shifted recently to lowering dietary intakes and achieving higher bioavailability of nutrients. As a result, there is a critical need for an appropriate delivery vehicle for active ingredients, as their fortification poses numerous issues, such as physical or chemical instability, low bioavailability, incompatibility with food matrix, and unpleasant taste. Nanoformulations are mostly synthesized using bottom-up or top-down processes. The challenge is frequently addressed by nanoformulations, which encapsulate the fortificant in an appropriate loading vehicle for fortification of the desired food matrix. The use of a nanoformulation-based delivery system with an appropriate design may overcome the limitations associated with food matrix fortification. Effective nanoformulation design and fabrication for active ingredient delivery in the food system provides advantages such as protection against biochemical and microbial degradation, retention of sensory attributes such as texture, flavor, mouthfeel, and overall acceptance; improved bioavailability and storage stability of fortified product.

Nevertheless, concerns about consumer health and safety are an ongoing challenge in dealing with the development of nanotechnology in food systems, and thus mandatory testing of nanofoods is required before they are released to the market. This chapter highlights the reported knowledge relevant to the use of *T. occidentalis* in food fortification and as a therapeutic agent based on its prominent biological activities and the presence of phytochemicals. There are no elaborate reports on the significance of nanofood fortification in enhancing conventional methods of food fortification. This study aims to bridge the gap so that the need for planned and well-structured experimental designs to deliver possible future applications could be highlighted. To our knowledge, this is the first article that unites two essential aspects, food application of *T. occidentalis* and nanoencapsulation as an approach for enhancing food fortification.

2. Popular and common/vernacular names

T. occidentalis is commonly known in the following languages and countries: Fluted pumpkin, oyster nut, oil nut, fluted gourd and Telfairia nut (English); Oroko, pondokoko and Gonugbe (Sierra Leone); Costillada (Spanish); *"iroko"* or *"apiroko"* (Yoruba-Nigeria), *"ubong"* (Efik-Nigeria), *"ugwu"* (Igbo-Nigeria), *"umeke"* (Edo-Nigeria), and *"umee"* (Urhobo-Nigeria); Krobonko (Ghana) [5, 14].

3. Description of the plant

T. occidentalis is a dioecious, perennial, tropical vine grown for its leaves and edible seeds [15]. It is a creeping herbaceous vegetable having lobed leaves and twisted tendrils which extends over the soil [16]. Fluted pumpkins can be grown on flat lands or on mounds. It is commonly grown beside fences or adjacent to a tree in domestic gardens, thus allowing the fruit to be suspended from a branch [17]. It can also be grown on a variety of trellis, including bamboo [5]. **Figure 1** shows a photograph of the plant (**Table 1**).

Figure 1.
Photo of Telfairia occidentalis *[18].*

4. Botanical classification of *Telfairia occidentalis*

Kingdom	Plantae–plantes, Planta, Vegetal, plants
Subkingdom	*Viridaeplantae*–green plants
Infrakingdom	*Streptophyta*–land plants
Division	*Tracheophyta*–vascular plants, *tracheophytes*
Subdivision	*Spermatophytina–spermatophytes*, seed plants, phanérogames
Infradivision	*Angiospermae*–flowering plants, angiosperms, plantas com flor, angiosperma, plantes à fleurs, angiospermes, plantes à fruits
Class	*Magnoliopsida*
Superorder	*Rosanae*
Order	*Cucurbitales*
Family	*Cucurbitaceae*–gourds, squashes, citrouilles, gourdes
Genus	*Telfairia* Hook.
Species	*Telfairia occidentalis* Hook. f.–oysternut.

Adopted from Eseyin et al. [19] and Imosemi [18].

Table 1.
Telfairia occidentalis *is classified as follows by the integrated taxonomic information system.*

5. Nutritional value of *T. occidentalis*

5.1 Leaves

T. occidentalis is a storehouse of important nutrients. The leaves of *T. occidentalis* [20] have a wide range of medicinal, industrial, and nutritional properties [19, 21]. The leaves, according to Akanbi et al. [22], are high in fat (18%), minerals, and vitamins (20%). *T. occidentalis* leaves have a greater protein content (56%) than most other green leafy vegetables [23]. The leaves are also a rich source of P, Ca, Zn, Fe, K, Cu, and Mn [2, 19, 24, 25] which are essential in human and animal nutrition [8]. *T. occidentalis* had a total amino acid content of 455.3 mg/g, with a total essential amino acid content of 256.1 mg/g, or 56.3%, indicating that the plant proteins are high in essential amino acids. The essential amino acid contents are compared favorably with those of important legumes [26]. The essential amino acid profile of *T. occidentalis* had also been shown to be very rich and include alanine, aspartate, glycine, glutamine, histidine, lysine, methionine tryptophan, cysteine, leucine, arginine, serine, threonine, phenylalanine, valine, tyrosine, and isoleucine [7, 27, 28]. According to Iweala and Obidoa's [29] study, long-term feeding of *T. occidentalis*-supplemented diet caused a significant increase in the weight of animals, which may be due to its rich nutrient content. Vitamins E and C are present in aqueous extracts of the leaf at 5.07 mg/100 mL and 40 mg/100 mL, respectively [19]. Palmitoleic acid (16.62%) and elaidic acid (0.85%) are the predominant omega-9 fatty acid present in the leaf [30]. *T. occidentalis* leaves are rich in beta-carotene and contained a significantly high amount of vitamin C, total flavonoids and phenolics than *Psidium guajava* stem bark. *T. occidentalis* is rich in minerals that are blood boosters such as iron, folic acid, copper, zinc, potassium, cobalt, sodium, and calcium; vitamins such as nicotinamide, thiamine, vitamins A, C, and K, and other core amino acids: glycine, aspartate, leucine, isoleucine, alanine, arginine, serine, methionine, tryptophan, phenylalanine, valine, tyrosine, cysteine, threonine, and histidine [28, 31, 32]. Phytochemicals such as alkaloids, tannins, terpenoids, glycosides, saponins, anthraquinones, reducing sugar, flavonoids, and phenolic compounds, are present in *T. occidentalis* leaves [2, 19].

5.2 Seeds

T. occidentalis seeds can be ground and added to soups, or they can be roasted, cooked, and eaten. The dried seeds contain 32.50% fat, 11.43% carbohydrates, 34.56% protein, 15.71% fibers, and 4.40% total ash [33]. Glucose, fructose, sucrose, and sixteen amino acids are also present, with glutamic acid (16.4 g/100 g) being the highest and lysine (2.6 g/100 g) being the lowest. In addition, phospholipid, glycolipid, and neutral lipid contents of 58, 26, and 15%, respectively, are contained in the seeds [34, 35]. The antioxidant activity of the seed has also been reported by Eseyin et al. [35] and Osukoya et al. [36]. The seed also contained significant amounts of vitamins A and C, which can be used to supplement other dietary sources. Unsaturated fatty acids make up 61% of fluted pumpkin seed oil [24]. The high content of unsaturated fatty acids in the seed confers a high nutritive value on these seeds. The younger seeds are nutritionally preferable as food because they contain fewer anti-nutrients and have a sweeter taste than the mature seeds [5]. *T. occidentalis* oil has high iodine values compared to palm oil, indicating that the oil has a high content of unsaturated fatty acids relative to palm oil. This suggests that it could be used as a cooking oil or in the production of margarine [37].

5.3 Anti-nutritive factors

Anti-nutritional factors are the biologically active secondary metabolites produced by plants as side products for their own defense and reduce the absorption of macronutrients (proteins) and micronutrients (vitamins and minerals) [38]. In *T. occidentalis*, phytic acid (22.11 mg/100 g) is present [2], which does not have any negative effect on the body if taken up to 10–60 mg/g but its prolonged intake can cause decreased bioavailability of some essential minerals in case of monogastric animals as phytate binds these nutrients in the digestive tract and causes their deficiency. It mainly decreases the levels of overall calcium, iron, and zinc balance [39]. The content of oxalate in *T. occidentalis* is (0.35 mg/100 g) [2] which have no any negative effect on the body as the established permissible levels of oxalate in the human body are 250 mg/100 g of food samples [40]. Soluble oxalates chelate calcium and magnesium that is released in the digestive system, making these micronutrients unavailable for absorption and utilization. Calcium is a mineral that the body requires for strong bone formation and maintenance. It is also involved in some hormonal and enzymatic functions, and in nerve impulse coordination. Soluble oxalates are excreted through the kidneys if absorbed, where they can also cause stone formation (calcium oxalate crystals). Oxalates that are insoluble have no metabolic function in the body and are excreted in feces [41]. Tannin content (4.98 mg/100 g) present in *T. occidentalis* has no bad effects on the health as it is acceptable up to 560 mg [42, 43]. However, its higher concentration of tannin in the diet minimize Fe absorption [44]. This component binds to Fe in the lumen, lowering Fe bioavailability, especially non-heme iron found in plant foods like leafy green vegetables [45]. They are harmful at high levels and interfere with protein digestion and absorption, as well as vitamin and mineral utilization [46, 47].

5.4 Processing of *T. occidentalis*

Different processing methods used also influence the contents of nutrients and anti-nutrients in *T. occidentalis*. Most plants lose their nutritive properties when processed. On comparing the nutritive content of boiled, roasted, and fermented *T. occidentalis* seed flour, it was found that all the nutrients, including protein, vitamins C, B3, and E, β-carotene, iron, and zinc, were significantly higher in the fermented samples [48]. The effects of roasting periods on the nutritional, anti-nutritional, and mineral compositions of *T. occidentalis* seeds were evaluated [49]. Interestingly, as a result of roasting *T. occidentalis* seeds for 60 minutes, there is a significant increase in all nutrients and mineral contents and decrease in the anti-nutrients [49]. In another study, Fluted pumpkin seeds were processed into raw, boiled, fermented, germinated, and roasted seeds, dried at 50°C, milled and sieved [50]. It was found that germination and fermentation enhance the protein quality of the fluted pumpkin seed flour. This can be a result of the biochemical activities during germination and microbial activity during fermentation, it also reduces deleterious elements and improve zinc bioavailability [50]. A study reported the anti-nutrient composition, including oxalate, phytic acid, cyanide, and tannin of *T. occidentalis* leaf, determined at three temperature regimes (normal (37°C), 60°C and boiling point (100°C) [51]. It was observed that the boiled sample at 100°C processing condition was most impactful for anti-nutrient reduction. In another research, Fagbemi et al. [52], reported that processing significantly reduced anti-nutritional factors of fluted pumpkin seed. Traditional processing methods using aqueous systems with boiling and cooking have been shown to reduce anti-nutrient levels in foods, particularly vegetables. Cooking effectively is thus recommended to reduce the concentrations of anti-nutrients in foods to levels that are permissible [53–55], while allowing consumers to benefit from the other phytochemicals that *T. occidentalis* contain. The cell wall of the vegetables is raptured during cooking and blanching, thereby releasing anti-nutritional factors into the blanching medium [51, 56–59].

5.5 *T. occidentalis* as food fortificant

Fortification is used interchangeably with the enrichment of staple food to add specific micronutrients to enhance the nutritional value of prepared foods. The fortificants must be easily obtainable, need to be well absorbed, must not interfere with the sensory attributes of fortified food, and must be cost-effective. It can be done in different forms such as mass, market-driven, and targeted fortification. Fortified food consumption must be adequate and sufficient for the targeted population. Whatever the goal of fortification, fortificants must be compatible with food characteristics and impart nutritional value to fortified foods while retaining their appearance and other organoleptic properties [60]. Consumers are most interested in a product's appearance, which is regarded as an important influencing factor in their decision to purchase it. Fortification of *T. occidentalis* can be significant to tackle nutrient deficiencies and malnutrition. The researchers indicate its applications in different kinds of foods as *chin-chin* [61], cassava pasta, [62], smoothies, [63], soups condiment (*Ogiri ugu*) [64, 65]. The foods are described here in detail.

5.6 Smoothies

T. occidentalis leaves powder has been incorporated at the rate of 1.5, 3.0, and 4.5% to the different ratios of smoothies [63]. The addition of the leaves significantly improves proteins, carbohydrates, and minerals, especially calcium and potassium in the smoothies. But while *T. occidentalis* leaves were able to significantly improve the crude protein content, the percentage inclusion should not be more than 1.5 [63].

5.7 *Chin-chin*

Chin-chin is a traditional Nigerian snack prepared by combining wheat flour, butter, milk and eggs into a stiff paste and then deep frying until golden brown [66]. *Chin-chin* can sometimes be baked instead of fried [67]. The long shelf life of chin-chin allows for large-scale production and distribution. Furthermore, good eating quality makes *chin-chin* appealing for fortification and other nutritional improvements. *Chin-chin* is a high-energy food, rich in carbohydrates and fat [68] but low in other nutrients such as protein, minerals, and vitamins [69]. Efforts have been made to improve the nutritional content of *chin-chin* by supplementing it with leafy greens. Different percentages *Telfairia occidentalis* leaves and Indian Spinach vegetable powder were incorporated into wheat flour to develop *chin-chin* and assess its proximate, mineral, and sensory acceptability [61]. The enriched *chin-chin* had an increase in protein, fiber, fat, potassium, magnesium, calcium, iron, and zinc concentrations, and was found to be acceptable by the consumers.

5.8 Bread

Wheat bread is widely accepted and consumed worldwide. Bread is a baked product made traditionally from wheat flour. It's high in carbohydrate but low in protein, vitamins, and minerals [68]. Attempts have been made to improve the nutritional content of bread by supplementing with flours like wheat and undefatted rice bran [68], wheat, maize, and orange-fleshed sweet potato [70], and moringa seed powder [71], vegetables leaf powder [72]. In addition, efforts have been made to promote the use of composite flours in which flour from locally grown high-protein legumes/oilseeds replace a portion of wheat flour for the production of high-protein composite breads [73], of which *Telfairia occidentalis* seed is one of them. The full-fat seeds of *Telfairia occidentalis* have about 27% protein and 54% fat while defatted seeds have about 71% crude protein and are valuable as a high-protein oilseed for human food in Nigeria [74]. Besides being boiled and eaten as a vegetable, the seeds of *Telfairia occidentalis* are sometimes processed into flour or fermented and used as a protein supplement, functional agent, or flavoring ingredient in a variety of local foods [75–77]. Because of its high-water absorption capacity, *Telfairia occidentalis* seed flour has been reported to have good potential for use in bakery products [78, 79]. An evaluation of the seed flour's functionality in bread making revealed that up to 10% of wheat flour could be replaced with fluted pumpkin seed flour to produce acceptable bread [80]. Several studies have shown that the nutritional quality of bread improved when wheat flour was supplemented with legume/oilseed flours [81–83]. According to Giami et al., [84] when wheat flour was replaced with 10% defatted fluted pumpkin seed flour, there was an increase of 80.8% in crude protein, 43.9% in calcium, 71.9% in potassium, and 63.0% in phosphorus contents of composite breads. Diets formulated with 5 or 10%

fluted pumpkin-substituted bread had significantly higher values for weight gain, protein efficiency ratio, and apparent and true digestibilities than diets formulated with 100% wheat flour bread, indicating an improvement of the nutritional quality of fluted pumpkin-substituted composite bread.

6. Cookies

The development of value-added products from fluted pumpkin seed had been suggested as a way to increase the possibility of expanding the seed's utilization in the tropics [78, 79]. One potential food application for fluted pumpkin seed flour (FPF) is its use in composite flours for the production of bakery products, such as bread and cookies. Efforts have been made to promote the use of composite flours in which flour from locally grown high-protein oilseeds and legumes replace a portion of wheat flour for production of high-protein composite bakery products [73]. Replacing wheat flour (WHF) with defatted fluted pumpkin seed flour (FPF) at levels of 0–25% was studied for its effect on the chemical, physical, sensory, and nutritional properties of cookies [85]. The study showed that wheat flour supplemented with defatted FPF at the 5–15% levels produced acceptable cookies with spread ratio, hardness, color, and flavor similar to the control (100% WHF) cookies. When WHF was replaced with 15% FPF, there was an increase of 84.6% in crude protein, 62.9% in calcium, 131.0% in potassium, and 61.6% in phosphorus contents of composite cookies. Also, diets based on composite cookies containing 15% pumpkin flour were nutritionally comparable with a diet based on casein, indicating that the underutilized high protein fluted pumpkin seeds available in tropical countries could be processed into value-added products and used to combat malnutrition [85].

7. Complementary food

Traditional weaning foods made from plant staples frequently fail to meet the nutritional needs of infants due to their stiff consistency and high volume, resulting in a low-cost filling meal that frequently lacks adequate nutrients [86]. They are therefore known to poorly support growth and development. Poor combination and formulation have partly contributed to the poor performance of traditional complementary foods. A number of researches [87, 88] in Nigeria have shown that a combination of cereals and legumes or tubers with vegetables and animal-sourced food rather than single diets, better-supported growth and development. A study to formulate complementary food from a blend of fluted pumpkin seeds and quality protein maize (QPM) has been done by Adedokun et al. [89]. The investigation revealed that combinations of fluted pumpkin seed flour and QPM increased the protein quality and chemical composition of the formulated diet.

8. Cassava pasta

Lawal et al. [62] evaluated the techno-functional and sensorial properties of cassava pasta, as influenced by incorporating fluted pumpkin leaf powder and the cultivar variation effect. Cassava pasta fortified with leaf powder at 5 and 10% incorporation levels reduced particle sizes, whereas yellow cassava flour had a larger particle size distribution than white cultivar. Pasta color was significantly influenced, as lightness values decreased as leaf powder concentration increased. Fluted pumpkin inclusion reduced the gelation capacity of the flour blends while increasing their water solubility, swelling power, and oil absorption capacities of the products. Interestingly, the addition of fluted pumpkin leaf powder reduced cooking time and gruel solid loss while increasing weight gain in the formulated pasta. With the addition of leaf powder, the hardness and pasting viscosities of the gel and pasta decreased by 12%, improving the textural properties of the cassava pasta. Pasting temperatures in the fluted pumpkin-fortified pasta were also lower than in the gluten-laden wheat pasta. Furthermore, yellow cassava products had significantly higher pasting viscosities than white cassava products, and cultivar variation affected the thermal properties of the food products significantly. Consumers' overall acceptance and likelihood of purchasing the novel pasta were modest (**Figure 2**).

Considering the views of several such fortifications, it is suggested that such addition can be done to other snacks as well. Addition of *T. occidentalis* to the snacks can add nutritive value to the snacks. However, further studies on *T. occidentalis* fortified snacks is required before bringing the commercialized product to the market.

1. Yellow cassava pasta

2. White cassava pasta

3. Yellow cassava-fluted pumpkin (5%) pasta

4. White cassava-fluted pumpkin (5%) pasta

5. Yellow cassava-fluted pumpkin (10%) pasta

6. White cassava-fluted pumpkin (10%) pasta

Figure 2.
Cassava pasta made with 0, 5, and 10 g 100 g^{-1} *fluted pumpkin (*Telfairia occidentalis*) leaf powder [62].*

8.1 Other food applications of *T. occidentalis*

8.1.1 Ogiri

The fluted pumpkin seeds are used for the production of *ogiri*. It is a fermented product used as condiments to flavor and soups. The pumpkin seed is high in protein, therefore, the *ogiri* is a nutritious product and beneficial for people who have a deficiency of protein. The process of *ogiri* preparation is still traditional and also the packaging of the *ogiri* is in the leaves so it makes *Ogiri* more traditional [65]. *Ogiri* is an alkaline fermented food condiment made from fluted pumpkin seeds (*Telfairia occidentalis*). It is used for the preparation of soups by an ethnic group, Igbo in Southeastern Nigeria. *Ogiri* (**Figure 3**) is prepared by manually dehulling fluted pumpkin seeds and wrapping them in blanched plantain or banana leaves before boiling them for 6–8 hours. The seeds are then fermented for four to six days near

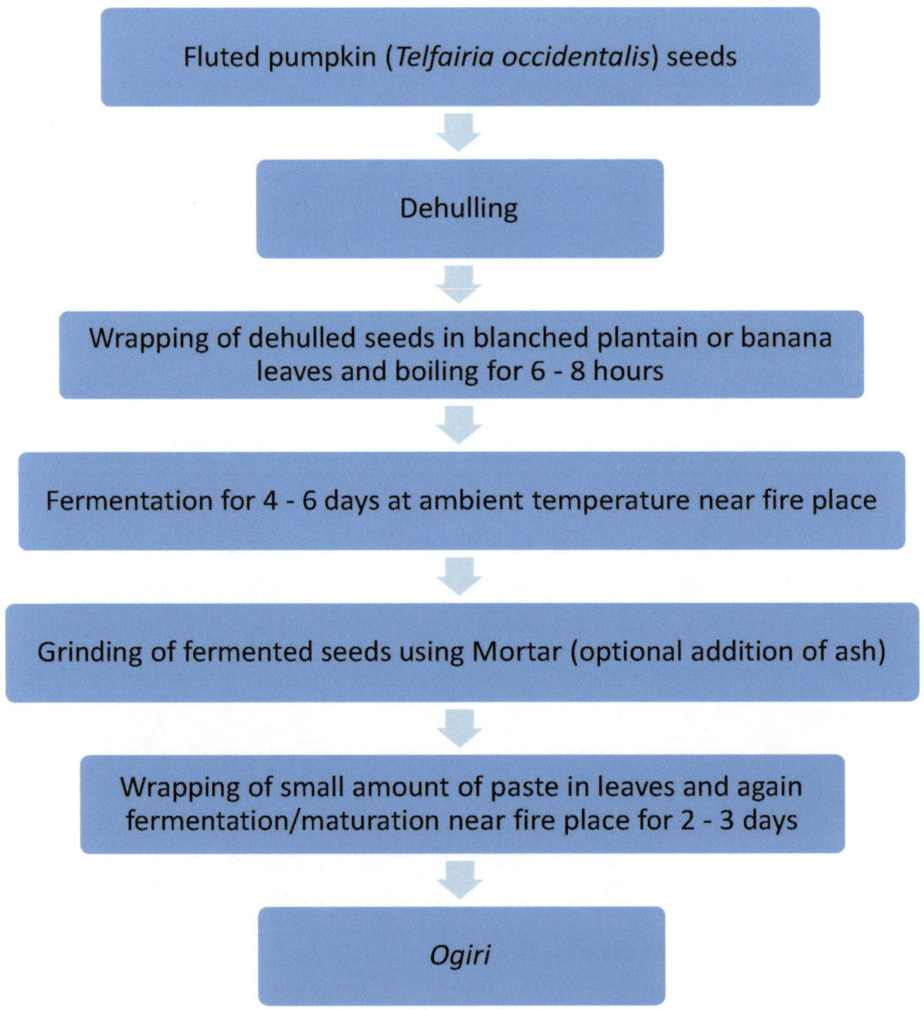

Figure 3.
Fermented T. occidentalis *seed (*Ogiri*) condiment wrapped with local leaves [64].*

a fireplace. After fermentation, cotyledons are sticky with a characteristic aroma. It is then ground into a fine paste with a mortar and pestle, and small portions of the paste are then wrapped in banana leaves and kept near the fireplace for further fermentation or maturation for two to three days. The pH of the product at the end of fermentation is found to be around 7.9 [90]. *B. subtilis*, *B. pumulis*, and *B. licheniformes* are the microorganisms responsible for fermentation (**Figure 3**).

8.2 Nanotechnology approaches for enhancing food fortification

In recent years, large groups of the population have become increasingly deficient in micronutrients such as vitamins and minerals. Hence, essential micronutrients are incorporated into common foods by food fortification. Bioactive compounds are those that have an effect on the body and include carotenoids, essential oils, antioxidants, and molecules that are widely incorporated into food to increase its nutritional and health properties. These nutrients are normally present in plant-based foods in small amounts but have a wide range of health benefits to the human body. Conventional methods of food fortification do not completely satisfy the functional requirements for bioactive compounds. They also have drawbacks of unsatisfactory flavor profiles, and poor stability and bioavailability. These drawbacks can be overcome with the utilization of nanodelivery systems. The various nanotechnology techniques that can be used for food fortification include nanosuspensions, nanoemulsions, nanoliposomes, and cyclodextrin carriers. Nanofood fortification has a wide range of advantages in the protection of phytochemicals by using an encapsulation technique and some of the micronutrients which are degraded rapidly or not properly absorbed by the body can also be aided using food fortification in the nanoscale [91].

8.3 Nanoencapsulation

The term nanoencapsulation refers to the use of nanometer-scale encapsulation with films, layers, and coverings. The encapsulation layer is clearly nanometer in size, forming a protective layer on the food or flavor molecules/ingredients [92]. Nanoencapsulation technologies have the potential to meet food industry challenges regarding the effective delivery of health functional ingredients and the controlled release of flavor compounds. Nanoencapsulation packs substances in miniature by using techniques such as nanocomposite, nanoemulsification, and nanostructuration and provides a final product. The functional ingredient is transported to the desired site of action *via* nanoencapsulation. They safeguard the functional ingredient against chemical or biological degradation during processing, storage, and use. They must be able to regulate the release of the functional ingredient. Finally, the delivery system must be compatible with the final product's physical-chemical and qualitative properties. Nanocarrier systems are typically carbohydrate, protein, or lipid-based [93].

Nanoencapsulation is accomplished through the use of nano capsules. They have several advantages, including ease of handling, improved stability, oxidation resistance, retention of volatile ingredients, taste making, moisture-triggered controlled release, pH-triggered controlled release, consecutive delivery of multiple active ingredients, flavor character change, long-lasting organoleptic perception, and enhanced bioavailability and efficacy. They are defined as nano vesicular systems with a typical core-shell structure in which the drug is confined to a reservoir or cavity surrounded by a polymer membrane or coating. The active substance can be present in the cavity as a liquid, solid, or molecular dispersion. Nano capsules are involved in the delivery of the desired component and entrapment of the odor and unwanted components in the food and thereby resulting in the preservation of the food [93]. In the biological system, nano capsules transport food supplements through the gastrointestinal tract, increasing the substance's bioavailability. The primary advantage of encapsulation is that it protects the hidden component, allowing it to be delivered precisely to the target even in adverse conditions. Nano capsules can be prepared in six different ways: (i) nano precipitation, (ii) emulsion diffusion, (iii) double emulsification, (iv) polymer coating, and (v) layer-by-layer.

8.4 Nanoemulsions

Nanoemulsions are dispersions of two immiscible phases (dispersed and continuous) with nanoscale particle diameters (< 200 nm). Nanoemulsions are thermodynamically unstable but kinetically stable emulsions that differ in particle diameter from conventional emulsions [94]. Depending on whether the dispersed phase is oil or water, nanoemulsions can be either oil-in-water (o/w) or water-in-oil (w/o). By encapsulating lipophilic vitamins or omega-3 in the oil core, oil-in-water nanoemulsions have great potential for protecting, stabilizing, and delivering lipophilic bioactive compounds. Significant research has been conducted to develop nanoemulsion-based delivery systems for these compounds [95].

The formation of emulsions requires a water phase, an oil phase, an emulsifier, and either mechanical or physiochemical energy [96, 97]. An oil-in-water nanoemulsion has a three-component core-shell structure: a lipophilic core containing a lipophilic bioactive compound, a hydrophilic shell (aqueous phase), and an amphiphilic interface containing the emulsifier/surfactant [96]. Emulsifiers are surface-active molecules that adsorb on the oil-water interface of newly formed droplets and reduce interfacial tension, resulting in smaller droplet size emulsions. Depending on the processing conditions and emulsion composition, different types of food grade surfactants and emulsifiers can be used to form nanoemulsions [98, 99]. Because of the presence of hydrophilic groups and hydrophobic moieties in their structure, higher molecular weight biopolymers such as amphiphilic proteins (e.g., casein, lactoferrin, b-lactoglobulin, protein isolates, whey proteins) and polysaccharides (gum arabic and modified starch) can act as good emulsifiers [98, 100, 101]. Small molecule surfactants (e.g., tween 20, 40, 80, and Span 80), phospholipids (e.g., soy lecithin), quillaja saponins, sucrose esters are amphiphilic molecules that consist of a hydrophilic head and lipophilic tail groups [96, 99]. Nevertheless, among various types of emulsifiers, natural emulsifiers have great potential for use in food applications since there is an increasing demand for "clean-label" products in the global food market [101].

In general, nanoemulsions are produced either by high-energy methods (microfluidics, high-pressure homogenizers, or ultrasound equipment) or low-energy methods (spontaneous emulsification and phase inversion temperature) [102].

8.5 Nanoliposomes

Nanoliposomes are vectors that are used in both the pharmaceutical and food industries. These lipid nanostructures are incorporated into food products during the manufacturing process, primarily to improve texture, flavors, and food preservation. Nanoliposomes are an intriguing type of carrier for bioactive molecules due to their

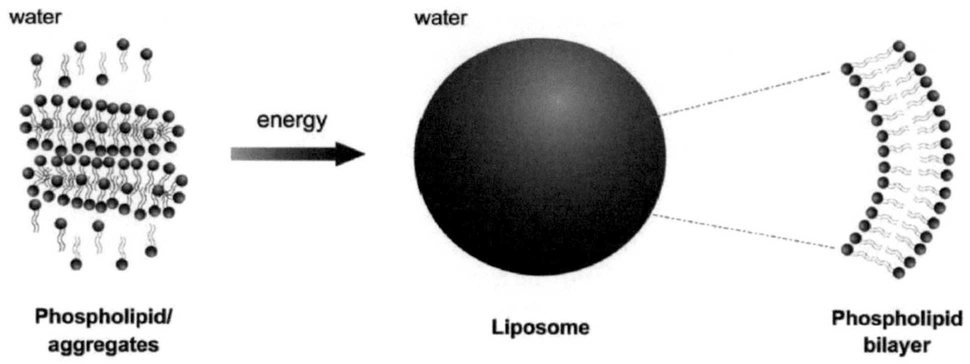

Figure 4.
A simplified mechanism for the formation of liposomes and nanoliposomes [104].

natural lipid composition and ability to encapsulate both hydrophobic and hydrophilic compounds. Encapsulation of molecules known for their beneficial effects on specific organs or tissues in these lipid-based vectors can be envisioned in nutraceutical applications to create functional foods designed for disease prevention. To achieve this goal, however, certain parameters must be controlled during the preparation and storage of nanoliposomes to ensure optimal digestibility and bioavailability. Indeed, challenges remain in ensuring the stability of nanoliposomes during storage as well as after ingestion. There are numerous preparation methods available, but the oxidative nature of lipids and their phase transition temperature all have an impact on the [103] stability of nanoliposomes (**Figure 4**) [105].

9. *T. occidentalis* as nutraceutical

The term nutraceuticals is derived from the words "nutrition" and "pharmaceuticals," and it refers to food-derived products that have health benefits. These have recently been investigated as suitable alternatives for the control and prevention of a wide range of diseases because they are considered safe and have potential nutritional value. Fruit and vegetables are good sources of functional foods because they contain a lot of phytochemicals, which have a lot of health benefits [19, 36, 106–117]. *T. occidentalis* and its various parts have been shown to have therapeutic and exploratory effects against a variety of diseases such as diabetes, cancer, malaria, and anemia [118, 119]. The presence of phytochemicals makes it a good medicinal agent.

9.1 Role of *T. occidentalis* in malnutrition

Malnutrition is a condition caused by eating a diet in which nutrients are either not enough or too much low which can cause health problems. It could be deficient in calories, carbohydrates, proteins, vitamins, or minerals and ultimately resulting in nutritional problems. Malnutrition casts long dimness, 800 million people are affected in which 20% of all people in the developing world [120]. It is noted from nutritional analysis that *T. occidentalis* leaves contain a wealth of essential, disease-preventing nutrients. Experimental finding has confirmed that *T. occidentalis* effectively maintain electrolyte balance, modulates pancytopenia and oxidative hepatorenal damage in rats suggesting its protective potential against anemia and organ failure [121, 122]. In another study, the efficacy of ethanol leaf extract of *T. occidentalis* in phenyl hydrazine model of anemia in rats was investigated by Oladele et al., [113]. They found treatment with ethanol leaf extract of *T. occidentalis* modulates all the anomalies such as pancytopenia, anemia, and other related disorders which are characterized by altered hematological, biochemical, and molecular indices, compromised cellular and structural integrity, as well as physiological functions, suggesting its ameliorating effects against phenyl hydrazine-induced anemia and hematotoxicity, and may be useful in the treatment of chemically induced anemia or other related diseases. Fasuyi and Nonyerem [31] documented that the inclusion levels of about 15% *T. occidentalis* leaf meal in broiler starter diets was found to be the most nutritionally suitable protein supplement in their diets.

10. Conclusion and future prospects

The leaves and seeds of *T. occidentalis* are rich sources of phenolic compounds, essential amino acids, vitamins, proteins, and β-carotenes, which can play a regulatory and functional role. *T. occidentalis* leaves are considered as more nutritive than the other parts of this plant. *T. occidentalis* leaf powder is used as dietary supplement and also protects humans against iron deficiency, malnutrition, diabetes, cancer, and oxidative stress. Furthermore, fortification with its leaf powder improves the nutritional, technological, and functional properties of baked products.

As the demand for snacks in the market is huge, hence future prospects should focus more on the fortification of *T. occidentalis* into snacks to eradicate malnutrition, which has a twin advantage. The phytochemical contents of *T. occidentalis*-fortified products also need to be determined in future research. Traditional food fortification methods do not completely meet the functional requirements for bioactive compounds. They also have unsatisfactory flavor profiles, as well as poor stability and bioavailability. Yet another focal area of research is the effective design and fabrication of nanoparticle-based delivery systems for fortification of desired food matrix, which provides benefits such as protection against biochemical and microbial degradation, retention of sensory attributes such as texture, flavor, mouthfeel, and overall acceptance; improved bioavailability and storage stability of the fortified product. The plant as a native to Nigeria can become a great source of income for the nation if this potential for highly nutritional food is exploited by the industries and researchers by undertaking further research to corroborate earlier studies. This could be a "lead" to the discovery of novel nano-fortified food products. It will also improve the proper utilization of this popular and vital plant's numerous benefits.

Author details

Aisha Idris Ali[1*], Munir Abba Dandago[1] and Fatima Idris Ali[2]

1 Department of Food Science and Technology, Kano University of Science and Technology, Wudil, Nigeria

2 Department of Biological Sciences, Bayero University, Kano, Nigeria

*Address all correspondence to: aishaidrisali@gmail.com

References

[1] Batool M, Ranjha MMAN, Roobab U, Manzoor MF, Farooq U, Nadeem HR, et al. Nutritional value, phytochemical potential, and therapeutic benefits of pumpkin (Cucurbita sp.). Plants. 2022;**11**:1394. DOI: 10.3390/plants11111394

[2] Idris AA. An interventional approach to combat micronutrient malnutrition through herbal food product prepared from fluted pumpkin (*Telfairia occidentalis*) leaves, curry (*Murraya koenigii*) leaves and spinach (*Spinacia oleracea*) leaves [Ph.D. thesis]. Prayagraj: Sam Higginbottom University of Agriculture Technology and Sciences; 2022. pp. 66-67

[3] Ahluwalia SS, Goyal D. Microbial and plant derived biomass for removal of heavy metals from wastewater. Bioresource Technology. 2007;**98**(12):2243-2257. DOI: 10.1016/j.biortech.2005.12.006

[4] Saini RK, Sivanesan I, Keum YS. Phytochemicals of *Moringa Oleifera*: A review of their nutritional, therapeutic and industrial significance. 3 Biotech. 2016;**6**(2):1-14. DOI: 10.1007/s13205-016-0526-3

[5] Akoroda MO. Ethnobotany of *Telfairia occidentalis* (*Cucurbitaceae*) among Igbos of Nigeria. Economic Botany. 1990a;**44**:29-39

[6] Badifu GI, Ogunsua AO. Chemical composition of kernels from some species of *Cucurbitaceae* grown in Nigeria. Plant Foods for Human Nutrition. 1991;**41**:35-44

[7] Kayode AA, Kayode OT. Some medicinal values of *Telfairia occidentalis*: A review. American Journal of Biochemistry and Molecular Biology. 2011;**1**:30-38

[8] Idris S. Compositional studies of *Telfairia occidentalis* leaves. American Journal of Chemistry. 2011;**1**:56-59

[9] Mojisola C-OC, Agbedahunsi JM, Akinola NO. Studies on the effect of a nutritious vegetable, *Telfairia occidentalis*, on HbSS blood. Journal of Traditional and Complementary Medicine. 2019;**9**:156-162

[10] Idris AA, Paul V, Chattree A. Evaluation of the use of different solvents for phytochemical constituents and antioxidants activity of the leaves of *Telfairia occidentalis* Hook (*Cucurbitaceae*). Juni Khyat (UGC Care Group I Listed Journal). 2021;**11**(01):200-208

[11] Airaodion AI, Ogbuagu EO, Ekenjoku JA, Ogbuagu U, Airaodion EO. Therapeutic effect of methanolic extract of *Telfairia occidentalis* leaves against acute ethanol-induced oxidative stress in wistar rats. International Journal of Bio-Science and Bio-Technology. 2019;**11**:7

[12] Akindele AJ, Oladimeji-Salami JA, Usuwah BA. Antinociceptive and anti-inflammatory activities of *Telfairia occidentalis* hydroethanolic leaf extract (*Cucurbitaceae*). Journal of Medicinal Food. 2015;**18**:1157-1163

[13] Nkereuwem AO, Eseyin OA, Udobre SA, Ebong A. Evaluation of antioxidant activity and chemical analysis of the leaf of *Telfairia occidentalis*. Nigerian Journal of Pharmaceutical and Applied Science Research. 2011;**1**(1):21-28

[14] Gbile ZO. Vernacular names of Nigerian plants (Yoruba). Ibadan:

Forestry Research Institute of Nigeria.-. En Geog; 1984. p. 101

[15] Phillip IB, Shittu AM, Aiyelaagbe OO, Adedokun T. Economic potentials of plantain and fluted pumpkin intercropping as a poverty reduction strategy in south-western Nigeria. World Journal of Agricultural Sciences. 2009;**5**:525-534

[16] Horsefall M, Spiff AI. Equilibrium sorption study Al3+, Co2+, Ag+ in aqueous solutions by fluted pumpkin (*Telfairia occidentalis* hook f) waste biomass. Acta Chimica Slovenica. 2005;**52**:174-181

[17] Nwauwa LO, Omonona BT. Efficiency of vegetable production under irrigation system in Ilorin metropolis: A case study of fluted pumpkin (*Telfairia occidentalis*). Continental Journal of Agricultural Economics. 2010;**4**:9-18

[18] Imosemi OI. Review of the toxicity, medicinal benefits, pharmacological actions and morphological effects of *Telfairia occidentalis* Hook. F. European Journal of Pharmaceutical and Medical Research. 2018;**5**(7):22-32

[19] Eseyin AO, Sattar MA, Rathore HA. A review of the pharmacological and biological activities of the aerial parts of *Telfairia occidentalis* Hook.f. (Cucurbitaceae). Tropical Journal of Pharmaceutical Research. 2014;**13**(10):1761-1769

[20] Kwenin WKJ, Wolli M, Dzomeku BM. Assessing the nutritional value of some African indigenous green leafy vegetables in Ghana. Journal of Animal and Plant Sciences. 2011;**10**(2):1300-1305

[21] Aworunse OS, Bello OA, Popoola JO, Obembe OO. Pharmacotherapeutic properties of *Telfairia occidentalis* Hook F.: A systematic review. Pharmacognosy Reviews. 2018;**12**:238-249

[22] Akanbi WB, Adeboye CO, Togun AO, Ogunrinde JO, Adeyeye SA. Growth, herbage and seed yield and quality of *Telfairia occidentalis* as influenced by cassava peel compost and mineral fertilizer. Agricultural Journal. 2007;**2**:588-595

[23] Mukhtar MA, Idris MB, Abdulrasheed A. The mineral composition and proximate analysis of *T. occidentalis* (fluted pumpkin) leaves consumed in Kano metropolis, Northern Nigeria. Chemical Science International Journal. 2016;**10**(1):1-4. DOI: 10.9734/ACSJ/2016/20632

[24] Eseyin OA, Ebong P, Ekpo A, Igboasoiyi A, Oforah E. Hypoglycemic effect of the seed extract of *Telfairia occidentalis* in rat. Pakistan Journal of Biological Sciences. 2007;**10**:498-501

[25] Gbadamosi SO, Famuwagun AA, Nnamezie AA. Effects of blanching with chemical preservatives on functional and antioxidant properties of fluted pumpkin (*Telferia occidentalis*) leaf. Nigerian Food Journal. 2018;**36**(1):45-57

[26] Aisegbu JE. Some biochemical evaluation of fluted pumpkin seeds. Journal of the Science of Food and Agriculture. 1987;**40**:151-155

[27] Fasuyi AO. Nutritional potentials of some tropical vegetable leaf meals: Chemical characterization and functional properties. African Journal of Biotechnology. 2006;**5**:49-53

[28] Tindall HD. Commercial Vegetable Growing. 1st ed. Oxford, UK: Oxford University Press; 1968

[29] Iweala E, Obidoa O. Some biochemical, haematological and histological responses to a long term consumption of *Telfairia occidentalis*-supplemented diet in rats. Pakistan Journal of Nutrition. 2009;**8**:1199-1203

[30] Inuwa HM, Aimola IA, Muhammad A, et al. Isolation and determination of omega-9 fatty acids from *Telfairia occidentalis*. International Journal of Food Nutrition and Safety. 2012;**1**:9-14

[31] Fasuyi AO, Nonyerem AD. Biochemical, nutritional and haematological implications of *Telfairia occidentalis* leaf meal as protein supplement in broiler starter diets. African Journal of Biotechnology. 2007;**6**(8):1055-1063. Available from: http://www.academicjournals.org/AJB

[32] Ajibade SR, Balogun MO, Afolabi OO, Kupolati MD. Sex differences in biochemical contents of *Telfairia occidentalis* Hook F. Journal of Food, Agriculture and Environment. 2006;**4**:155-156

[33] Chuku LC, Chinaka NC. Proximate and Phytochemical Compositions of Dried Seeds of Fluted Pumpkin (*Telfairia occidentalis*). Journal of Complementary and Alternative Medical Research. 2021;**13**(1):14-19. Article no. JOCAMR.64445. ISSN: 2456-6276

[34] Daramola OO, Oyeyemi WA, Odiase LO, Olorunfemi AA. Effects of methanol extract of *Telfairia occidentalis* seed on ovary antioxidant enzymes, serum hormone concentration and histology in wistar rats. International Journal of Pharmacognosy and Phytochemical Research. 2016;**8**:1245-1249

[35] Eseyin AO, Daniel A, Paul TS, Attih E, Emmanuel E, Ekarika J, et al. Phytochemical analysis and antioxidant activity of the seed of *Telfairia occidentalis* Hook (*Cucurbitaceae*). Natural Product Research. 2017. pp. 1-4. DOI: 10.1080/14786419.2017.1308366

[36] Osukoya OA, Adegbenro D, Onikanni SA, Ojo OA, Onasanya A. Antinociceptive and antioxidant activities of the methanolic extract of *Telfairia occidentalis* seeds. Ancient Science of Life. Oct-Dec 2016;**36**(2):98-103. DOI: 10.4103/asl.ASL_142_16. PMID: 28446831; PMCID: PMC5382825

[37] Christian A. Fluted pumpkin (*Telfairia occidentalis* Hook.f.) seed. A nutritional assessment. Electronic Journal of Environmental, Agricultural and Food Chemistry. 2007;**6**:1787-1793

[38] Yvonne K, Chidewe C, Zvidzai CJ. Phytochemical and anti-nutrient composite from selected marginalized Zimbabwean edible insects and vegetables. Journal of Agriculture and Food Research. 2020;**2**:100027

[39] Elinge CM, Muhammed M, Atiku FA, Itodo AU, Peni IJ, Sanni OM, et al. Proximate, Mineral and Anti Nutrient Composition of Pumpkin seed extract. International Journal of Plant Research. 2012;**2**(5):146-150

[40] Oguchi Y, Weerakkody WAP, Tanaka A, Nakazawa S, Ando T. Varietal differences of quality-related compounds in leaves and petioles of spinach grown at two locations. Bulletin of the Horishima Prefectural Agriculture Research Center. 1996;**64**:1-9

[41] Bong WC, Savage G. Oxalate content of raw, wok-fried, and juice made from bitter gourd fruits. Food Science & Nutrition. 23 Oct 2018;**6**(8):2015-2019. DOI: 10.1002/fsn3.706. PMID: 30510702; PMCID: PMC626120

[42] Gemede HF, Ratta N. Antinutritional factors in plant foods: Potential

health benefits and adverse effects. International Journal of Nutrition and Food Sciences. 2014;**3**(4):284-289

[43] Gemede HF, Ratta N. Anti-dietary factors in plant foods: Potential health benefits and adverse effects. Advanced Research Journal of Microbiology. 2018;**5**(2):100-113

[44] Afsana K, Kazuki S, Satoshi I, Hiroshi H. Reducing effect of ingesting tannic acid on the absorption of iron, but not of zinc, copper and manganese by rats. Bioscience, Biotechnology, and Biochemistry. 2004;**68**(3):584-592

[45] Disler PB, Lynch SR, Torrance JD, Sayers MH, Bothwell TH, Charlton RW. The mechanism of the inhibition of iron absorption by tea. The South African Journal of Medical Sciences. 1975;**40**(4):109-116

[46] Getachew A, Asfaw Z, Singh V, Woldu Z, Baidu-Forson JJ, Bhattacharya S. Dietary values of wild and semi-wild edible plants in Southern Ethiopia. African Journal of Food, Agriculture, Nutrition and Development. 2013;**13**(2):7485

[47] USEPA. Inert Reassessment - Tannin (CAS Reg. NO. 1401-55-4). Office of prevention, pesticide and toxic substance. Washington, D.C.: United States Environmental Protection Agency; 2006. p. 16. Available from: http://www.epa. gov/opprd001/inerts/tannin.pdf [Retrieved: October 02, 2014]

[48] Nzeagwu OC, Amadi AU. Effect of processing on chemical composition of fluted pumpkin (*Telfairia occidentalis*) seed flour and sensory evaluation of soup produced from the flour. Nigerian Journal of Nutritional Sciences. 2020;**41**(1)

[49] Miteu GD, Ezeh BC. Effects of roasting periods on the nutritive value of *Telfaira occidentalis* (fluted pumpkin) seeds. IPS Journal of Nutrition and Food Science. 2022;**1**(1):6-10. DOI: 10.54117/ijnfs.v1i1.2

[50] Fagbemi TN. Effects of processing on the nutritional composition of fluted pumpkin (*Telfairia occidentalis)* seed flour. Nigerian Food Journal. 2007;**25**(1):1-22

[51] Njoku PC, Nzediegwu E, Ayuk AA, Nzediegwu C, Efenudu IU, Erhayimwen MA. Anti-nutrient composition of pumpkin leaf (*Telfiaria occidentalis*) at three temperature regimes. Pakistan Journal of Nutrition. 2014;**13**:678-682

[52] Fagbemi TNF, Eleyinmi AF, Atum HN, Akpambang O. Nutritional composition of fermented fluted pumpkin (*Telfairia occidentalis*) seeds for production of "ogiri ugu". IFT Annual Meeting, July 15-20. New Orleans, Louisana; 2005

[53] Adefegha SA, Oboh G. Cooking enhances the antioxidant properties of some tropical green leafy vegetables. African Journal of Biotechnology. 2011;**10**(4):632-639

[54] Akhtar N, Ihsan-ul-Haq, Bushra M. Phytochemical analysis and comprehensive evaluation of antimicrobial and antioxidant properties of 61 medicinal plant species. Arabian Journal of Chemistry. 2018;**11**:1223-1235

[55] Mugova S, Sachs PR. Opportunities and Pitfalls of CSR: A Summary, Opportunities, and Pitfalls of Corporate Social Responsibility. Springer; 2019. pp. 241-246

[56] Kunatsa Y, Cathrine C, Cuthbert JZ. Phytochemical and anti-nutrient composite from selected

marginalized Zimbabwean edible insects and vegetables. Journal of Agriculture and Food Research. 2020;**2**(100027):1-7

[57] Natesh HN, Abbey L, Asiedu SK. An overview of nutritional and anti-nutritional factors in green leafy vegetables. Horticulture International Journal. 2017;**1**(2):58-65

[58] Zaini NA, Mustaffa F. Review: Parkia speciosa as a valuable, miracle of nature. Asian Journal of Medicine and Health. 2017;**2**(3):1-9

[59] Oboh G. Effect of blanching on the antioxidant properties of some tropical green leafy vegetables. LWT—Food. Science and Technology. 2005;**38**:513-517

[60] Sandeep G, Anitha T, Vijayalatha K, Sadasakthi A. Moringa for nutritional security (*Moringa Oleifera* Lam.). International Journal of Botany Studies. 2019;**4**:21-24

[61] Olubukola A, Saka OG, Kehinde AT, Durodoluwa JO. Proximate, mineral, sensory evaluations and shelf stability of chinchin enriched with Ugu and Indian Spinach Vegetables. International Journal of Biochemistry Research & Review. 2017;**18**(4):1-14

[62] Lawal OM, Sanni O, Oluwamukomi M, Fogliano V, Linnemann AR. The addition of fluted pumpkin (*Telfairia occidentalis*) leaf powder improves the techno-functional properties of cassava pasta. Food Structure. 2021;**30**(100241):1-11

[63] Aderinola TA. Effects of pumpkin leaves on the chemical composition and antioxidant properties of smoothies. In: Proceedings of the 4th Regional Food Science and Technology Summit (Refosts), 6-7 June 2018, Akure, Ondo state Nigeria. 2018

[64] Dimejesi SA. Aspect of microbiology and amino acid composition of "ogiri" Produced from Seeds of *Telfairia occidentalis*. International Digital Organization for Scientific Research. IDOSR Journal of Scientific Research. 2020;**5**(1):58-67

[65] Ogueke CC, Okoli AI, Owuamanam CI, Ikechukwu AP, Iwoun JO. Production of Soup Condiment (*Ogiri ugu*) from Fluted Pumpkin Seeds Using Bacillus subtilis. International Journal of Life Sciences. 2013;**2**(3):113-120

[66] Adegunwa MO, Ganiyu AA, Bakare HA, Adebowale AA. Quality evaluation of composite millet-wheat chinchin. Agriculture and Biology Journal of North America. 2014;**5**(1):33-39

[67] Akubor PI. Protein contents, physical and sensory properties of Nigerian snack foods (cake, chinchin and puff-puff) prepared from cowpea-wheat flour blends. International Journal of Food Science & Technology. 2004;**39**(4):419-424

[68] Ameh MO, Gernah DI, Igbabul BD. Physico-chemical and sensory evaluation of wheat bread supplemented with stabilized undefatted rice bran.Food Science & Nutrition. 2013;**4**:43-48

[69] Young J. Functional bakery products: current directions and future opportunities. Food Industrial Journal. 2001;**4**:136-144

[70] Igbabul B, Num G, Amove J. Quality evaluation of composite bread produced from wheat, maize and orange fleshed sweet potato flours. American Journal of Food Technology. 2014;**2**(4):109-115

[71] Bolarinwa IF, Aruna TE, Raji AO. Nutritive value and acceptability of bread fortified with moringa seed

powder. Journal of the Saudi Society of Agricultural Sciences. 2019;**18**:195-200

[72] Odunlade TV, Famuwagun AA, Taiwo KA, Gbadamosi SO, Oyedele DJ, Adebooye OC. Chemical composition and quality characteristics of wheat bread supplemented with leafy vegetable powders. Journal of Food Quality. 2017;**2017**:7

[73] UNECA (United Nations Economic Commission for Africa). Technical Compendium on Composite Flours. Ethiopia: United Nations Economic Commission for Africa; 1985

[74] Longe OG, Farinu GO, Fetuga BL. Nutritional value of the fluted pumpkin (*Telfairia occidentalis*). Journal of Agricultural and Food Chemistry. 1983;**31**:989-992

[75] Achinewhu SC. Protein quality evaluation of weaning food mixtures from indigenous fermented foods. Nigerian Journal of Nutritional Sciences. 1987;**8**:23-31

[76] Banigo EB, Akpapunam MA. Physico-chemical and nutritional evaluation of protein–enriched fermented maize flour. Nigerian Food Journal. 1987;**5**:30-36

[77] Barber LI, Ibiama EA, AchinewhuSC.Microorganismsassociated with fermented fluted pumpkin seeds (*Telfairia occidentalis*). International Journal of Food Science and Technology. 1989;**24**:189-193

[78] Giami SY, Isichei I. Preparation and properties of flours and protein concentrates from raw, fermented and germinated fluted pumpkin (*Telfairia occidentalis* Hook) seeds. Plant Foods for Human Nutrition. 1999;**54**:67-77

[79] Giami SY, Bekebain DA. Proximate composition and functional properties of raw and processed full-fat fluted pumpkin (*Telfairia occidentalis*) seed flour. Journal of the Science of Food and Agriculture. 1992;**59**:321-325

[80] Giami SY. Effect of addition of fluted pumpkin (*Telfairia occidentalis*) seed flour on rheological properties of dough and baking quality of bread. In: Nkama I, Jideani VA, Ayo JA, editors. Proceedings of the 24th Annual Conference of the Nigerian Institute of Food Science and Technology, 20-24 November, Bauchi, Nigeria. 2000. pp. 29-30

[81] Kailasapatty K, Perera PAJ, Macneil JH. Improved nutritional value in wheat bread by fortification with full-fat winged bean flour (*Psophocarpus tetragonolobus* L.D.C). Journal of Food Science. 1985;**50**:1693-1696

[82] Nmorka GO, Okezie BO. Nutritional quality of winged bean composite breads. Cereal Chemistry. 1983;**60**:198-202

[83] Nochera C, Caldwell M. Nutritional evaluation of breadfruit–containing composite flour products. Journal of Food Science. 1992;**57**:1420-1422 1451

[84] Giami SY. Nutritional evaluation of germinated fluted pumpkin (*Telfairia occidentalis* Hook) seeds. Plant Foods for Human Nutrition. 2003;**58**:1-9. DOI: 10.1023/B:QUAL.0000040313.58273.E5

[85] Giami SY, Achinewhu SC, Ibaakee C. The quality and sensory attributes of cookies supplemented with fluted pumpkin (*Telfairia occidentalis* Hook) seed flour. International Journal of Food Science & Technology. 2005;**40**:613-620

[86] Fernandez DP et al. Fatty acid, amino acid and trace mineral analyses of five weaning foods from Jos, Nigeria. Plant Foods for Human Nutrition. 2002;**57**:257-274

[87] Fashakin JB, Ogunsola E. The utilization of local foods in formulation of weaning foods. Journal of Tropical Pediatrics (London). 1982;**28**:93-96

[88] Oyeleke AO et al. The use of cowpea (Vigna unguiculata) in improving a popular Nigeria weaning food. British Journal of Nutrition. 1985;**54**:343

[89] Adedokun OA, Fagbemi TN, Osundahunsi TO, Oboh G. Evaluation of Complementary Food from Blend of Fluted Pumpkin (*Telfairia occidentalis*) and Quality Protein Maize. Acta Scientific Nutritional Health. 2020;**4**(1). ISSN: 2582-1423

[90] Omafuvbe BO, Falade OS, Osuntogun BA, Adewusi SRA. Chemical and biochemical changes in African Locust Bean (Parkia biglobosa) and Melon (Citrullus vulgaris) seeds during fermentation to condiments. Pakistan Journal of Nutrition. 2004:140-145. DOI: 10.3923/pjn.2004.140.145

[91] Abinash V, Rahul T, Antoniraj MG, Moses JA, Anandharamakrishnan C. Nanotechnology approaches for food fortification. In: Pal K, Banerjee I, Maji S, et al, editors. Food, Medical, and Environmental Applications of Polysaccharides. Elsevier; 2021. pp. 161-186. DOI: 10.1016/C2018-0-05112-9. ISBN: 978-0-12-819239-9

[92] Paredes Javier A, Mariana Asencio C, Juan Manuel L, Alberto Allemandi D, Palma SD. Nano encapsulation in the food industry: Manufacture, applications and characterization. Journal of Food Bioengineering and Nanoprocessing. 2016;**1**(1):56-79. Available from: https://www.researchgate.net/publication/297683856_Nanoencapsulation_in_the_food_industry_manufacture_applications_and_characterization

[93] Amra B, Jasmin S. Micro- and nano-encapsulation in food industry. Croatian Journal of Food Science and Technology. 2019;**11**(1):113-121

[94] McClements DJ. Nanoemulsions versus microemulsions: Terminology, differences, and similarities. Soft Matter. 2012;**8**:1719-1729

[95] McClements DJ. Nanoemulsion-based oral delivery systems for lipophilic bioactive components: Nutraceuticals and pharmaceuticals. Therapeutic Delivery. 2013;**4**:841-857

[96] McClements DJ, Rao J. Food-grade nanoemulsions: Formulation, fabrication, properties, performance, biological fate, and potential toxicity. Critical Reviews in Food Science and Nutrition. 2011;**51**:285-330

[97] Tadros T, Izquierdo P, Esquena J, Solans C. Formation and stability of nano-emulsions. Advances in Colloid and Interface Science. 2004;**108-109**:303-318

[98] Chanamai R, McClements DJ. Comparison of gum arabic, modified starch, and whey protein isolate as emulsifiers: Influence of pH, CaCl2 and temperature. Journal of Food Science. 2002;**67**:120-125

[99] Kralova I, Sj€oblom J. Surfactants used in food industry: A review. Journal of Dispersion Science and Technology. 2009;**30**:1363-1383

[100] Adjonu R, Doran G, Torley P, Agboola S. Whey protein peptides as components of nanoemulsions: A review of emulsifying and biological functionalities. Journal of Food Engineering. 2014;**122**:15-27

[101] Ozturk B, McClements DJ. Progress in natural emulsifiers for utilization in

food emulsions. Current Opinion in Food Science. 2016;**7**:1-6

[102] Tansu Ö, Turhan S. Physicochemical properties of pumpkin (Cucurbita pepo L.) seed kernel flour and its utilization in beef meatballs as a fat replacer and functional ingredient. Journal of Food Processing and Preservation. 2020;**44**(9):e14695. DOI: 10.1111/jfpp.14695

[103] Khorasani S, Danaei M, Mozafari MR. Nanoliposome technology for the food and nutraceutical industries. Trends in Food Science & Technology. 2018;**79**:106-115

[104] Mozafari MR, Johnson C, Hatziantoniou S, Demetzos C. Nanoliposomes and their applications in food nanotechnology. Journal of Liposome Research. 2008;**18**(4):309-327

[105] Bondu C, Yen FT. Nanoliposomes, from food industry to nutraceuticals: Interests and uses. Innovative Food Science & Emerging Technologies. 2022;**81**:103140

[106] Akang EN, Oremosu AA, Osinubi AA, Dosumu OO, Kusemiju TO, Adelakun SA, et al. Histomorphometric studies of the effects of *Telfairia occidentalis* on alcohol-inducedgonado-toxicity in male rats. Toxicology Reports. 2015;**2**:968-975

[107] Jimoh TO. Enzymes inhibitory and radical scavenging potentials of two selected tropical vegetable (*Moringa oleifera* and *Telfairia occidentalis*) leaves relevant to type 2 diabetes mellitus. Revista Brasileira de Farmacognosia. 2018;**28**:73-79

[108] Kayode AA, Kayode OT, Odetola AA. *Telfairia occidentalis* ameliorates oxidative brain damage in malnorished rats. International Journal of Biological Chemistry. 2010;**4**:10-18

[109] Kayode OT, Kayode AA, Odetola AA. Therapeutic effect of *Telfairia occidentialis* on protein energy malnutrition-induced liver damage. Research Journal of Medicinal Plants. 2009;**3**:80-92

[110] Kambizi L, Bakare-Odunola MT, Oladiji AT, Kola-Mustapha AT, Amusa TO, Atolani O, et al. Proteinease inhibition, membrane stabilization, antioxidant and phytochemical evaluations of leaves, seeds and calyces of four selected edible medicinal plants. Cogent Chemistry. 2017;**3**(1):1314064. DOI: 10.1080/23312009.2017.1314064

[111] Nwanna EE, Oboh G. Antioxidant and hepatoprotective properties of polyphenol extracts from *Telfairia occidentalis* (fluted pumpkin) leaves on acetaminophen induced liver damage. Pakistan Journal of Biological Sciences. 2007a;**10**:2682-2687

[112] Oyewole OA, Abalaka ME. Antimicrobial activities of *Telfairia occidentalis* (fluted pumpkins) leaf extract against selected intestinal pathogens. Journal of Health Science. 2012;**2**:1-4

[113] Oladele OJ, Oyeleke OM, Awosanya OO, Olowookere BD, Oladele OT. Fluted pumpkin (*Telfairia occidentalis*) protects against phenyl hydrazine-induced anaemia and associated toxicities in rats. Advances in Traditional Medicine. Springer; 2020. DOI: 10.1007/s13596-020-00499-7

[114] Oboh G, Nwanna E, Elusiyan C. Antioxidant and antimicrobial properties of *Telfairia occidentalis* (fluted pumpkin) leaf extracts. Journal of Pharmacology and Toxicology. 2010b;**5**:539-553

[115] Oboh G, Rocha JB. Antioxidants in Foods: A New Challenge for Processors: Leading Edge Antioxidants Research. New York: Nova Science Publishers Inc; 2007

[116] Obi RK, Nwanebu FC, Ndubuisi-Nnaji UU, Onuoha LN, Chiegboka N. Ethanolic extraction and phytochemical screening of two Nigerian herbs on pathogens isolated from wound infections. International Journal of Comprehensive Pharmacy. 2011;**10**:1-5

[117] Eze BC, Ogbodo EC, Ezejindu DN, Ezeugwunne IP, Analike RA, Onuora IJ, et al. Hepatoprotective potential of the aqueous leaf extract of *Telfairia occidentalis* on the Liver function parameters in Adult Wistar Rats. Annals of Geriatric Education and Medical Sciences. 2020;**7**(1):39-42

[118] Rodriguez-Casado A. The health potential of fruits and vegetable phytochemicals: Notable examples. Critical Reviews in Food Science and Nutrition. 2016;**56**(7):1097-1107

[119] Vanitha A, Vijayakumar S, Ranjitha V, Kalimuthu K. Phytochemical screening and antimicrobial activity of wild and tissue cultured plant extracts of Tylophoraindica. Asian Journal of Pharmacy and Pharmacology. 2019;**5**(1):21-32

[120] Dutta D, Heo I, Clevers H. Disease modeling in stem cell-derived 3D organoid systems. Trends in Molecular Medicine. 2017;**23**(5):393-410. DOI: 10.1016/j.molmed.2017.02.007

[121] Oladele JO, Oyewole OI, Bello OK, Oladele OT. Hepatoprotective effect of aqueous extract of *Telfairia occidentalis* on cadmium chloride-induced oxidative stress and hepatotoxicity in rats. Journal of Drug Design and Medicinal Chemistry. 2017a;**3**:32-36

[122] Oladele JO, Oyewole OI, Bello OK, Oladele OT. Modulatory properties of *Telfairia occidentalis* leaf extract on pancytopenia, electrolyte imbalance and renal oxidative damage in rats. Journal of Bioscience and Biotechnology Discovery. 2017b;**2**:74-78

Chapter 6

Indigenous South African Food: Nutrition and Health Benefits

Samkeliso Takaidza

Abstract

Many populations around the world rely on indigenous plant materials as their primary source of nutrition. On the continent of Africa, many rural communities continue to gather, grow, and consume these food crops. In South Africa, there are numerous indigenous food crops, such as cereals, green vegetables, and various kinds of wild fruit. This review discusses the advantages of particular indigenous foods for nutrition and health. Current literature shows that indigenous foods possess vital macronutrients and have positive health benefits. Indigenous crops have the potential to provide options for long-term food security because they have been found to be more resistant to climate change. They might also present opportunities for new products. Literature also indicates that the benefits and value of these traditional foods are still not completely appreciated in South Africa and throughout Africa. The biggest obstacle to achieving these societal benefits is the lack of scientific information about the nutritional content, health benefits, efficient processing, and preserving technology of indigenous food. The potential value of the indigenous food system may be improved if its advantages were more properly investigated.

Keywords: bioactive compounds, minerals, indigenous, nutrition, health, phytocompounds

1. Introduction

Indigenous plants have long been used by humans as food and medicine in practically all communities. They grow naturally in ecosystems and are known to be potent sources of nutrients, therefore better for health. Forager and hunter-gatherer societies historically relied only on local flora for their diets [1]. Currently, to combat food insecurity, many people rely on a varied diet that includes edible plant material harvested from the wild as well as edible plant material grown in backyard gardens. For instance, on the African continent, more than 95% of households use grass, roots, and leaves from wild plants to supplement their diets. To maintain food self-sufficiency and the continued existence of millions of people, these communities cultivate and consume native fruits and vegetables [2]. It has been suggested that indigenous foods can significantly improve diet quality, and food and nutrition security. Despite this claim, the consumption of indigenous foods has decreased as a result of these foods' unavailability in contemporary industrialized and commercialized markets and a lack of investment in research and development. Consumption, therefore, is associated with poverty and low self-esteem among rural black people [1, 3].

In South Africa, there is a wide variety of locally grown food as well as collected in the wild that is not only nutritious but also effective in preventing chronic illness [4]. Indigenous plant meals have long played a significant role in folklore. Many people in rural areas still harvest and eat both cultivated foods and wild vegetation [2]. There is evidence from studies conducted in South Africa and around the world, that indigenous food is consumed by many households, especially in areas where they are available, and thus can play an important role in alleviating food insecurity and contribute to dietary diversity. These include indigenous grain crops, vegetables, and fruits [1, 2].

Indigenous grain crops produce seeds that are abundant in protein and high in starch. Cereals and pulses are two categories for these crops. Pulses include cowpea and Bambara while cereals include pearl millet and sorghum. Indigenous vegetable crops that can be utilized for food preparation include those with tender leaves, stems, and stalks. These include root/tubers like amadumbe and leafy vegetables such as cleome, cowpea, amaranth, and blackjack. Indigenous fruit crops are fleshy, seed-associated structures that are palatable and delicious when uncooked. These fruits include marula, monkey oranges, and sour figs, as examples [5].

This chapter, therefore, describes these indigenous crops, and their nutritional and health benefits. The diverse indigenous plant foods in South Africa may be exploited to develop ideas and items for many markets, such as new and natural colorants and flavors, medications, and dietary supplements.

2. Indigenous South African food: nutrition and health benefits

Food crops classified as indigenous to South Africa include those that were domesticated there as well as those that were brought in and are now regarded as traditional crops. These crops are grown and produced in a variety of climates, with many of them being found in the wild. Grains, vegetables, and fruit make up the three categories of indigenous crops [5].

3. Indigenous grain crops

Millet was a significant food crop in the distant past in Africa, where it is said to have originated. The introduction of maize was likely one of the main factors in Southern Africa, in which millet's role was diminished to that of a small cereal crop with limited economic relevance [6]. Recently there has been improvement in the cultivation of millet in South Africa with Limpopo, KwaZulu Natal, and Free state provinces being the major producers [7, 8].

3.1 *Pennisetum glaucum* (Pearl millet)

Pearl millet (*Pennisetum glaucum*), a grass of the Poaceae family, is one of the most significant cereal crops grown in the tropics and in developing nations (**Figure 1**). On about 27 million acres, pearl millet is grown in some of the harsh, tropical areas of sub-Saharan Africa and Asia. It is a crop with numerous applications, such as for food, fuel, and livestock, however, as food, it is primarily used for thin and thick

Figure 1.
Pearl millet crop (A) and grains (B) [9].

porridge. Smallholder farmers in Southern Africa have grown Pearl millet largely for subsistence, but its uses have expanded over time [10].

Pearl millet grains are beneficial due to their high fiber, fatty acid, and phytochemical content. It has a high mineral content, 70% carbohydrate content, 9–13% protein, 5–7% lipids, 70–74% polyunsaturated fatty acid, and high mineral content (Fe, Zn, Ca, Mg and P) [11]. Pearl millet's anti-inflammatory, antihypertensive, anticarcinogenic, and antioxidant qualities assist prevent heart disease, inflammatory bowel disease, and atherosclerosis. Pearl millet grains are naturally gluten-free, which is beneficial because wheat's gluten protein has been related to metabolic disorders such as intestinal permeability, allergies, intolerances, and autoimmune diseases. Pearl millet may be a low-cost alternative for celiacs, people with non-celiac gluten sensitivity (NCGS), and gluten-sensitive people [12]. As a source of soluble and insoluble fiber, it increases insulin sensitivity, reducing the risk of type II diabetes [11]. Pearl millet's anti-nutritional components (phytate, tannins, and oxalic acid) restrict its use [13]. Therefore, anti-nutrients must be removed. Decortication, heating, soaking, germination, and fermentation may reduce antinutrient levels [14].

3.2 *Sorghum bicolor* L. (Sorghum)

Sorghum (*Sorghum bicolor*, L.), (**Figure 2**), which was initially cultivated in Northeast Africa more than 60 centuries ago, is a drought-tolerant and climate-resistant crop. Most of the world's sorghum is grown in Africa (39.2%), the Americas (38.2%), and Asia (18.3%) [16]. Sorghum is a crop of the Gramineae family that is high in carbohydrates. It was domesticated between 3000 and 5000 years ago, and today, after wheat, maize, rice, and barley, it is thought to be the fifth most significant carbohydrate-rich crop in the world. Sorghum plays a key role in the fight against hunger and food insecurity [17].

Sorghum is traditionally prepared by boiling, roasting, baking, and deep-frying. Unfermented or fermented sorghum grains can form sorghum porridge (typically combined with other grains like maize and millet). It is also used to make alcohol [18]. Sorghum contains slowly digesting starch, low-digestibility proteins (kafirins), and unsaturated fatty acids. It has phosphorus, potassium, zinc, and B-complex vitamins thiamine, riboflavin, and pyridoxine (D, E, and K). Red, dark, and black cultivars

Figure 2.
Sorghum crop (A) and grain (B) [15].

include 3-deoxy anthocyanidins and tannins [17]. Sorghum is now used in breakfast cereals, infant formula, confectionery, and gluten-free baking. The restricted usage of sorghum in the food sector is partly due to challenges in processing sorghum endosperm, where significant protein cross-linking of hydrophobic kafirins hinders starch functioning during hydrothermal cooking. Sorghum is unique among main cereal grains since it possesses significant quantities of bioactive components that can enhance human health. Polyphenols, notable flavonoids, are sorghum's most useful bioactive components. Sorghum is abundant in bioactive lipids (particularly policosanols and phytosterols) in the grain pericarp waxes. Sorghum endosperm has a slower starch digesting profile than other cereal grains, which modulates postprandial blood glucose response. Research on sorghum's ability to reduce chronic diseases has focused on polyphenols and endosperm digestion. Some types of cancer can be prevented, glucose metabolism is enhanced, insulin resistance is lowered, lipid metabolism is improved, fat storage is repressed, and gut flora is modified to promote colon health, as examples of positive impacts [19]. In vitro studies demonstrate that sorghum compounds, especially 3-deoxy anthocyanidins, tannins, and lipids, modulate noncommunicable disease and gut microbiota functions (obesity, diabetes, dyslipidemia, cardiovascular disease, cancer, and hypertension). Understanding sorghum's bioactive compounds that support these benefits is crucial [17].

3.3 *Vigna unguiculate* (Cowpea)

Cowpea (*Vigna unguiculata* L. Walp), is an indigenous African legume crop, having originated in Africa and spread throughout Latin America and Southeast Asia [20] Cowpeas often grow as climbing or trailing vines with three-lobed compound leaves. At the ends of long stalks, the white, purple, or pale-yellow flowers typically appear in pairs or clusters of three as in **Figure 3**. Depending on the cultivar, the pods can reach a length of 20–30 cm (8–12 inches) and are long and cylindrical [22].

Cowpea is a member of the Fabaceae family, subtribe Phaeseolinae, Vigna genus, and Catjang section. The *V. unguiculata* subspecies unguiculata is responsible for all cultivated cowpeas (**Figure 4**). The black-eyed pea, black-eyed bean, Crowder pea, Southern pea, frijol caup, and feijo-caup are all names for this legume crop [24]. The cowpea plant is predominantly cultivated for its grain (dry or fresh) and leaves. Cowpea seed flour is versatile and important for making several dishes and snacks. For example,

Figure 3.
Cowpea leaves and pods [21].

Figure 4.
Six seed coat color patterns in cowpea: (A) Black color; (B) Blackeye color (C) Browneye color; (D) Cream color; (E) Pinkeye color; and (F) Red color [23].

cowpea seeds can be consumed raw, cooked, parched, fried, roasted, combined with the sauce, or stewed. Cowpeas are full of nutrients that are good for health, including fiber, plant-based proteins, vitamins (A, C, thiamine, riboflavin, folate, and B6), iron, selenium, zinc, magnesium, phosphorus, and copper. Black-eyed peas' high fiber content and plant-based proteins lower hunger hormones and facilitate weight loss [23]. However, cowpea possesses a lot of antinutrients such as phytic acid and protease inhibitors. Numerous treatment techniques, such as soaking, germination, fermentation, or debranning, can be beneficial to increase the availability of nutrients in cowpeas and reduce their antinutrient characteristics.

Cowpeas improve heart health, blood circulation, and cholesterol levels. Cowpeas' low glycaemic index reduces blood sugar spikes. As a result, adding them to the diet is good for diabetics. Cowpeas boost collagen synthesis, which improves brightness, reduces wrinkles and restores skin. Black-eyed peas are rich in folate, which is helpful for pregnant women and fetal growth [25]. Cowpea leaves and green pods are used traditionally to cure measles, smallpox, adenitis, burns, and ulcers. Cowpea seeds are astringent, antipyretic, and diuretic. The decoction is useful for liver and spleen difficulties, intestinal pains, leucorrhoea, monthly abnormalities, and urine expulsions. Cowpea flatulence hinders its consumption. The presence of raffinose, a stomach-upsetting fiber, may induce abdominal pain, gas, and bloating in some persons. Soaking or fermenting cowpeas before cooking reduces flatulence and improves nutrient absorption [26].

3.4 *Vigna subterranean* (Bambara groundnut)

The Bambara groundnut (*Vigna subterranea* L. *Verd*c.; Syn: *Voandzeia* subterranean L. Thouars.) is an indigenous, underutilized legume species widely produced in drier areas of sub-Saharan Africa, including South Africa [27]. It is considered the third most common major legume after groundnuts (*Arachis hypogea*) and cowpeas (*Vigna unguiculata*) on the African continent. South Africans grow Bambara groundnuts for subsistence. It is an annual plant with creeping, low-growing branches (**Figure 5A**). It takes three to six months to mature, depending on the variety and weather. After pollination, pods grow underground. Unripe pods are yellowish-green and have up to six pods, whereas mature pods may be purple (**Figure 5B**). The 1–5 cm long pod is spherical or oval. It sometimes has two seeds. Mature pods are indehiscent and yellow to reddish-dark brown [29]. Its dried seed can be eaten like any other dry vegetable or used to make soy-like vegetable milk. The seeds are eaten alone or with maize. Bambara nut is a nutrient-rich legume

Figure 5.
Bambara groundnut mature plant (A) and pods (B), [28].

considered a "balanced diet." Dried Bambara seeds contain 64.4% carbohydrates, 23.6% protein, 6.5% fat, and 5.5% fiber and are rich in micronutrients such as potassium (11.44–19.35 mg/100 g), iron (4.9–48 mg/100 g), sodium (2.9–12.0 mg/100 g), and calcium (95.8–99 mg/100 g) [30].

Bambara groundnut also has different physical and medicinal benefits; for example, the water from steamed seeds of Bambara groundnut is used to remedy diarrhea. Bambara groundnut diets may contain nutraceuticals that protect against high blood pressure and oxidative stress. The sap from Bambara groundnut leaves can be administered to the eyes to cure epilepsy. South African pregnant women consume raw seeds to manage nausea and vomiting [31]. Flavonoids and tannins are found in Bambara groundnut seeds, according to Tan et al. [32] and these compounds are common in dark or crimson seed coatings. Total phenolics and seed coat darkness are correlated. Despite the favorable health effects linked to the ingestion of Bambara groundnut, their antinutritional effects should not be disregarded.

4. Traditional leafy vegetables

Leafy vegetables include succulent young stems, blossoms, and fruits. Traditional African groups eat green vegetables as a traditional practice. Green leaves of indigenous plants are consumed fresh or cooked and eaten with porridge as a relish (spinach, morogo, imfino) [33]. *Amaranthus, Cleome, Corchorus*, and *Vigna* leaves can be short-boiled, sun-dried, and kept for further use or trading. Most of the green vegetables consumed by rural South Africans are weeds or wild plants. Local sourcing limits availability, quantity, and time. In South Africa's rural areas, more species could be cultivated, but little is known about their agronomic needs [34].

4.1 Amaranthus

The family Amaranthaceae, which has around 70 distinct species, includes the genus *Amaranthus*. Amaranth species, such as *A. doubtful* L., *A. hybridus*, and *A. tricolor* L., are often found in Asia and Africa and are used as leafy vegetables [35]. In South Africa, various amaranth species are used. Some are native, while others are weeds that have spread naturally from Europe and the Americas. One of the most popular leafy vegetable species is *A. hybridus* L (**Figure 6**) [5]. The young leaves, growth tips, and whole seedlings are gathered from fallow pastures and fields. Amaranth is harvested, prepared, and served as a side dish to corn porridge. Since amaranth is believed to naturally grow there, along with many other traditional leafy vegetables, it is rarely planted in South Africa. However, in the Bushbuckridge region of the provinces of Limpopo and Mpumalanga in South Africa, women do collect and store seed, which is then disseminated on their field when a decline in the natural population is noted. Another technique used to replenish natural seed stores is selective weeding, which protects amaranth and other traditional green crops growing in cultivated fields [36].

Amaranthus contains considerable amounts of dietary minerals, vitamins, proteins, and bioactive substances that may be beneficial to health [35]. It has historically been used as an analgesic, to induce labor, as an antipyretic, laxative, diuretic, digestible, anti-diabetic, anti-snake venom, and as a therapy for jaundice. Its immune-modulating, anti-inflammatory, antibacterial, and antioxidant antimalarial effects have all been documented [37].

Figure 6.
Amaranthus hybridus plant [5].

4.2 *Bidens pilosa* (black jack)

Bidens is a flowering plant genus with over 230 species belonging to the Asteraceae family. It is commonly called by numerous names such as hairy beggar tick; Spanish needles; devil needles; blackjack [38]. This genus is distinguished by the number of thorny bristles that stick onto animal coats or human garments. The number of thorns ranges from two to four; some species possess lengthy ray flowers and serrated segments, while others have undivided lance-shaped leaves with short ray flowers or none at all, while the majority have yellow disk flowers (**Figure 7**). *Bidens pilosa* is also used as a traditional leafy vegetable and to improve human health. The young shoot tips are used in tea and juice. The potential of *Biden Pilosa* for food and as a source of supplement is supported by its nutritional content, which includes calcium, phosphorus, percentage moisture, energy, and carotene [40, 41].

All parts of *B. pilosa* plant have been employed globally in traditional medicine to treat a variety of diseases, with specific indications differing from nation to nation. *Bidens pilosa* is a significant traditional medicine in South Africa that has been utilized for a variety of treatments by numerous cultural groups. For example, a leaf decoction is used to treat flatulence, ear infections, headaches, and renal issues. Additionally, the whole plant is employed as a poison antidote, and the leaf extract is utilized to treat malaria, stomach and mouth ulcers, diarrhea, hangovers, and hangovers [38, 42]. There are several main chemical compounds (about 300 constituents) belonging to phytosterols, fatty acids, pheophytins, terpenes, phenolic acids, okanin glycosides, chalcones, aurones, flavone glycosides, flavonoids, polyacetylene glycosides, and polyacetylenes identified and isolated from different parts of blackjack plant. The plant's therapeutic effects appear to be attributed to its bioactive phytochemical components, particularly sesquiterpene lactones and polyacetylenes, which inhibit the growth of pathogenic microorganisms, and flavonoids, which are recognized as effective anti-inflammatory agents. It has been discovered that the

Figure 7.
Bidens pilosa, A: flowers, B: Leaves C: Fruit [39].

phytochemicals and essential oil of *Bidens pilosa* contain amounts of phenolic compounds that can scavenge or neutralize free radicals [42]. Despite its usefulness, *B. pilosa*'s application is hindered by its categorization as inversive species [41].

4.3 *Cleome gynandra* (Spider plant)

The *Cleome*, a member of the Cleomaceae family is indigenous to South Africa. The tender leaves or young branches, and frequently the flowers (**Figure 8**), are boiled and eaten as a potherb, pleasant condiment, stew, or side dish throughout Africa. Other mashed foods incorporate fresh leaves as an ingredient while weaning foods contain dried leaves that have been ground and added. Due to *Cleome*'s bitterness, other green vegetables including cowpea (*Vigna* spp.), amaranth (*Amaranthus* spp.), and blackjack (*B. pilosa*) are often cooked with these leaves. *Cleome* is a good source of nutrients, including minerals (calcium and Iron) and vitamins (A and C). Leaves can be ground up and used to make a medicine that is taken internally to treat conditions like scurvy. In some cultures, leaves are prepared as a healthful meal that is believed to improve vision, provide energy, and alleviate marasmus. They are then boiled and marinated in sour milk for two to three days. It is a dish that pregnant and nursing moms are strongly warned against eating [44].

Cleome is highly regarded for its varied medicinal benefits against a number of ailments and conditions in addition to its nutritious composition and features.

Figure 8.
Traditional leafy vegetable, Cleome gynandra, growing at Genoa village in the Limpopo Province, South Africa. (A) Whole plant; (B) leaves; (C) flowers [43].

The majority of a plant's parts, including the leaves, are utilized to treat a variety of illnesses. In several nations, a decoction of *Cleome gynandra* leaves is used to treat malaria. Many pain-related diseases, including headache, neuralgia, stomachache, earache, rheumatoid arthritis, skeletal fractures, colic discomfort, and chest pain, are treated using the leaves and blossoms of *Cleome gynandra*. *Cleome gynandra* may contain anti-inflammatory phytochemicals that help reduce both acute and chronic inflammation, according to research on the treatment of pain-related illnesses. The idea that leaves contains a range of antibacterial chemicals is supported by their usage in the treatment and management of wounds, abscesses, diarrhea, and chancroid [45].

4.4 *Corchorus olitorius* (Wild orkra)

Corchorus olitorius, often known as wild okra or jute mallow, is a member of the Tiliceae family. It is a tropical, fast-growing, annual herb, up to 4 m tall, with a fibrous stem and yellow flowers in 1 to 2 months from the time of germination (**Figure 9**).

Corchorus olitorius is one of the most commonly consumed traditional African vegetables. It is obtained from both farmed and wild sources and is rich in vitamins, mineral salts, and folic acid, making it an essential dietary component. The leaves are consumed as part of vegetable-based side dishes served with starchy foods. It can be prepared alone, in combination with legume leaves or other wild vegetables, or with fish sauce. Therefore, it could play a crucial role in Africa's food security and fight against poverty [47]. *Corchorus olitorius* grows in the wild in KwaZulu Natal, Mpumalanga, Limpopo, and Gauteng, South Africa. It is primarily consumed in Limpopo, where it is gathered from the wild and eaten either fresh or dried. It is known to have high concentrations of iron and folate, which are beneficial for preventing anemia. In folk medicine, wild okra is used to cure gonorrhea, chronic cystitis, pain, fever, and tumors [48].

Figure 9.
Images of Corchorus olitorius in different habitats [46].

Figure 10.
Colocasia esculenta plant (a), leaves (b), and corm (c) [50].

4.5 *Colocasia esculenta* (L) Schott (Amadumbe)

Colocasia esculenta (L.) Schott; also known as taro is an annual herbaceous plant from tropical and subtropical climates and a member of the Araceae superfamily [49]. *Colocasia esculenta* is a tall herb that blooms and bears leaves simultaneously (**Figure 10**). It may be tuberous or have a thick, short caudex. Stem above ground zero, or slightly inflated at the base of the leaf sheaths, emerging from a firm, tapering rhizome, or, in cultivated varieties, a tuberous rhizome, with occasionally present suckers and stolons [51].

Important nutrients include carbohydrates, protein, thiamine, riboflavin, niacin, oxalic acid, calcium oxalate, minerals, lipids, unsaturated fatty acids, and anthocyanins present in the corms of this plant. When compared to potatoes, sweet potatoes, cassava, and rice, taro is more nutritious. The simple digestion of its starch granules and its nonallergenic characteristics are other factors that make taro a valuable food plant. Taro has also historically been utilized as a medicinal herb for therapeutic purposes. Traditionally the corm of taro is used as a remedy for body aches. The juice extracted from corm is used in alopecia, as an expectorant, stimulant, appetizer, and astringent. All plant components of this species can be used to extract a wide range of bioactive chemicals [49].

In terms of amount and variety, the edible part of taro is a rich source of antioxidants, primarily phenolic chemicals. Taro phytochemicals exhibit antimicrobial, immunomodulatory, antioxidant, anticancer, antimetastatic, antimutagenic, anti-hyperglycemic, and anti-hypercholesterolemic bioactivities in addition to their antioxidant properties. The use of taro, a potential alternative staple food with a lower glycemic index than potatoes, may reduce the incidence and prevalence of a number of diseases [52, 53].

5. Indigenous fruits

The majority of indigenous fruits are still picked from wild trees in many areas of Southern Africa. Previous research demonstrated that indigenous fruits and their by-products are abundant in phytochemicals with antibacterial and antioxidant activities, such as phenolic acids, flavonoids, triterpenes, and lignans. Phytochemical extracts from native fruits have been used as treatments for various illnesses, including sinusitis, fever, asthma, diarrhea, indigestion, and skin conditions [54].

5.1 *Sclerocarya birrea* (Marula)

Sclerocarya birrea (A. Rich.) Hochst, more commonly known as marula, is taxonomically derived from the Anacardiaceae plant family [55]. It is an indigenous, fruit-bearing tree of sub-Saharan Africa. Significant attention has been paid to the domestication and commercialization of *Sclerocarya birrea*. The fruits have stiff skin that ripens to a pale-yellow color and are plum-sized (3–4 cm in diameter) (**Figure 11**). Its flesh has a fibrous, mucilaginous texture, is extremely fragrant, and tastes juicy, sweet, and sour. The fruit's interior is a firm nut with 1–4 locules that encloses a soft, white seed that is rich in oil [54].

Traditionally, marula has multiple uses. Fruit pulp is widely consumed by humans as well as animals. Apart from raw consumption, the fruit pulp is used to make a variety of juices, jams, jellies, and liquor (Amarula) which are sold in national and international markets. The fruit pulp has been reported to contain high levels of

Figure 11.
Marula fruit [56].

vitamin C (54–194 mg/100 g). In addition, minerals such as calcium, iron, and zinc have been reported to occur in fruit pulp.

The seed kernel is an important source of edible oil. Monounsaturated fatty acids and natural antioxidants are abundant in marula oil. It can be categorized as having a high oleic acid content (70–78%) and a low tocopherol level. Therefore, it has been hypothesized that its fatty acid makeup accounts for its extraordinary stability. Recent research has suggested that a few of the oil's minor ingredients might potentially be contributing to this vital antioxidant function. Olive oil and marula oil both have similar fatty acid compositions, however, marula oil is 10 times more resistant to oxidation [57].

In addition, the seed kernel is used as a substitute for groundnuts in cooked vegetables, especially in South Africa. Other nutritional components such as fiber and carbohydrates have also been reported in *S. birrea* seed kernel. Various parts of *S. birrea* tree such as the root, leaves, bark, and seed kernel have, for a long time, been used for medicinal purposes. Extracts from the tree parts have acted as traditional remedies for treating diarrhea, headache, toothache, stomachache, swollen legs, anemia, malaria, high blood pressure, and scurvy. Several scientific examinations have also confirmed the availability of phytochemicals such as alkaloids, saponins, terpenoids, and tannins from the extracts of *S. birrea* tree parts. Thus, apart from consumption, fruit is also heavily exploited in the pharmaceutical industry [58].

5.2 *Strychnos spinosa* (Monkey orange)

The indigenous fruit tree *S. spinosa* has more than 30 aliases. The plant belongs to the Loganiaceae family, which contains more than 200 species in its largest genus. One of the common names for this fruit, "Spiny Monkey Orange," refers to the tree's spines and the fact that monkeys eat and seek out this fruit in great quantities [45].

The tree bears edible fruits which have a bright green wood peel (3–4 mm thick), which turns yellow–brown upon ripening (**Figure 12**). The fruit has an edible, juicy, sweet–sour pulp, which is pale brown in color and contains many hard brown (1–3 cm) seeds.

Strychnos spinosa trees provide tasty, high-nutrient fruits. The fruit is rich in carbohydrates, protein, lipids, fiber, and minerals. Monkey orange pulp contains 0.46 mg/100 g Cu, 0.23 mg/100 g thiamin, and 1.39 mg/100 g nicotinic acid [58]. Monkey orange pulp's sweet–sour flavor boosts consumer acceptance. The fruit includes 4 g/100 g of fiber, 70–140 mg of iron, and 34 mg of vitamin C. Micronutrients are important for development, bone formation, enzyme activity, and energy metabolism [60]. Fruits produce sweet juice, pulps, purees, and nectars, depending on the extraction process. *Strychnos spinosa* is KwaZulu-most Natal's (one of the provinces in South Africa) important indigenous fruit tree. Traditional production of fruit juice from *S. spinosa* species and enhanced technologies could benefit cooperatives and small-scale processors by broadening what they sell. The plant is used in traditional medicine to cure STDs, skin illnesses, snakebites, hypertension, malaria, pain, and inflammation. Different ethnic groups may use different plant components, preparations, and administrations to treat these conditions [61] More than 25 compounds have been profiled using gas chromatography–mass spectrometry (GC–MS), while about 45 compounds have been extracted and structurally characterized using UV–visible, Infrared (IR), Nuclear Magnetic Resonance (NMR), and mass spectrometry (MS). *Strychnos spinosa* has biological action against microorganisms (bacteria and fungi) and parasites (plasmodia, trypanosomes, and ticks) that cause human and animal sickness [45].

Figure 12.
Strychnos spinosa fruit- Unripe fruit (a), Mature fruit (b), Ripe edible fruit (c) [59].

Figure 13.
Hottentont fig plant (A) and fruits (B) [63].

5.3 *Carpobrotus edulis* (Hottentot-Fig)

Carpobrotus edulis L. (syn. *Mesembryanthemum edule* L., Aizoaceae), globally recognized as an aggressive invasive species of coastal environments is a medicinal and edible succulent extremophile plant native to the coast of South Africa. Various common names for it include sour fig, Cape fig, and Hottentots fig. In South Africa, *Carpobrotus edulis* has a long history of use in both traditional medicine and food [62].

The fruit produced are yellow, fleshy, spin-top-shaped fruits that turn reddish-brown, wrinkled, and leathery as they develop (**Figure 13**) [54]. Fresh, boiled, or dried *C. edulis* fruit can be used in pickles or chutney. *C. acinaciformis* and *C. delicious* have sweeter fruits, whereas *C. edulis* is astringent, salty, and sour. *Carpobrotus edulis* has been found to have high quantities of moisture (77.6–90.3%), carbohydrate (58.8–70.3%), energy (1240–1370 kJ 100 g1), and protein (8.1–26.0%). This data shows that *Carpobrotus* species could contribute to human nutritional needs as a dietary supplement since they contain a higher variety of important components than most commercially farmed fruits [64]. Leaf juice is used to cure diarrhea, TB, sore throats, gum infections, and burns. Eastern Cape traditional healers use it to treat intestinal worms, constipation, high blood pressure, and diabetes [62, 65, 66]. Fruit extracts alleviate skin and respiratory diseases, hypertension, and diabetes [54].

6. Conclusion

Indigenous foods provide nutritional and therapeutic benefits. Most of these crops are undervalued because they are considered weeds or poor man's food, lowering their economic position. Indigenous foods are gaining popularity due to the need to promote food security and provide alternate diets for those with diseases such as Type II diabetes. These meals' nutritional components and bioactive substances are determined using advanced techniques. Some of these foods, like cowpea, have been shown to have significant nutrition content but are limited by antinutritional factors. More research and funding should be dedicated to indigenous crop research to optimize their use.

Acknowledgements

The author acknowledges the Vaal University of Technology for the financial support.

Conflict of interest

The author declares no conflict of interest.

Author details

Samkeliso Takaidza
Vaal University of Technology, Vanderbijlpark, South Africa

*Address all correspondence to: parobek@tuzvo.sk

References

[1] Mbenyane XG. Indigenous foods and their contribution to nutrient requirements. South African Journal of Clinical Nutrition. 2017;**30**(4):5-7. Available from: http://www.sajcn.co.za/index.php/SAJCN/article/view/1299

[2] Rankoana SA. Indigenous plant foods of Dikgale community in South Africa. Journal of Ethnic Foods. 2021;**8**:5. DOI: doi.org/10.1186/s42779-021-00080

[3] Cloete PC, Idsardi EF. Consumption of indigenous and traditional food crops: Perceptions and realities from South Africa. Agroecology and Sustainable Food Systems. 2013;**37**:902-914. DOI: 10.1080/21683565.2013.805179

[4] Mngoma S. Let's get back to eating some traditional veggies. Foodformzansi.co.za. The new face of South African Agriculture. 2021. Available from: https://www.foodformzansi.co.za/why-little-understood-indigenous-foods-are-great-for-your-health/

[5] Louw M. Guide to indigenous food crops in South Africa. Available from: Southafrica.ac.za/guide-to-indigenous-food-crop.html

[6] De Wit JP, Chweigart FS. The potential role of pearl millet as a food in South Africa. S. -A. M E DIES E TYDSKRI F 21 Maart (Byvoegsel-Sllid-Afrikaanse Tydskrif vir Voedillg) National Nutrition Research Institute. In: Councilfor Scientific and Industrial Research. Pretoria; 1970

[7] Department: Agriculture, Forestry, and Fisheries (DAFF). Pearl millet Pennisetum glaucum. Available from: https://www.dalrrd.gov.za/Portals/0/Brochures%20and%20Production%20guidelines/PEARL%20MILLET%20BROCHURE%202014.pdf [Accessed: December 15, 2022]

[8] Graybosch RA. Encyclopedia of food grain. In: Wringley C, Corke H, Seetharama K, Faubion J, editors. The World of Food Grains. Second ed. Oxford: Elservier; 2016. pp. 16-21

[9] Green Millet. Available from: https://www.indiamart.com/proddetail/green-millet-22566076662.html [Accessed: December 15, 2022]

[10] Adebiyi JA, Obadina AO, Adebo OA, Kayitesi E. Comparison of nutritional quality and sensory acceptability of biscuits obtained from native, fermented, and malted pearl millet (*Pennisetumglaucum*) flour. Food Chemistry. 2017;**232**:210-217. DOI: 10.1016/j.foodchem.2017.04.020

[11] Krishnan V, Verma P, Saha S, Singh B, Vinutha T, Kumar RR, et al. Polyphenol-enriched extract from pearl millet (*Pennisetum glaucum*) inhibits key enzymes involved in post prandial hyper glycemia (α-amylase, α-glucosidase) and regulates hepatic glucose uptake. Biocatalysis and Agricultural Biotechnology. 2022;**43**:1-11. DOI: 10.1016/j.bcab.2022.102411

[12] Dias-Martinsa AM, Pessanha KLF, Pacheco S, Rodrigues JAS, Carvalho CWP. Potential use of pearl millet (*Pennisetum glaucum* (L.) R. Br.) in Brazil: Food security, processing, health benefits and nutritional products. Food Research International. 2018;**109**:175-186. DOI: 10.1016/j.foodres.2018.04.023

[13] Bello OM, Ogbesejana AB, Balkisu A, Osibemhe M, Musa B, Oguntoye SO. Polyphenolic fractions from three millet types *(Fonio, Finger millet, and*

Pearl millet): Their characterization and biological importance. Clinical Complementary Medicine and Pharmacology. 2022;**2**(1):1-10. DOI: 10.1016/j.ccmp.2022.100020

[14] Samtiyan M, Soni K, Chawla S, Poonia A, Sehgal S, Dhewa T. Key anti-nutrients of millet and their reduction strategies: An overview. Acta Scientific Nutritional Health. 2021;**5**:12

[15] Available from: https://www.britannica.com/plant/sorghum-grain [Accessed: December 10, 2022]

[16] George TT, Obilana AO, Oyenihi AB, Obilana AB, Akamo DO, Awika JM. Trends and progress in sorghum research over two decades, and implications to global food security. South African Journal of Botany. 2022;**151**:960-969. DOI: 10.1016/j.sajb.2022.11.025

[17] De Morais Cardoso L, Pinheiro SS, Martino HSD, Pinheiro-Sant'ana HP. Sorghum (*Sorghum bicolor* L.): Nutrients, bioactive compounds, and potential impact on human health. Critical Reviews in Food Science and Nutrition. 2017;**57**(2):372-390. DOI: 10.1080/10408398.2014.887057

[18] Meena AK, Visarada KBRS, Meena DK. Sorghum bicolor (L.) Moench a multifarious crop -fodder to therapeutic potential and biotechnological applications: A future food for the millennium. Future Foods. 2022;**2022**:1-14. DOI: 10.1016/j.fufo.2022.100188

[19] Girard ALAND, Awika JM. Sorghum polyphenols and other bioactive components as functional and health promoting food ingredients. Journal of Cereal Science. 2018;**2018**(84):112-124. DOI: 10.1016/j.jcs.2018.10.009

[20] Kritzinger Q, Lall N, Aveling TAS. (2005). Antimicrobial activity of cowpea (*Vigna unguiculata*) leaf extracts. South African Journal of Botany. 2005;**71**(1):45-48. DOI: 10.1016/S0254-6299(15)30147-2

[21] Available from: https://www.tropicalforages.info/text/entities/vigna_unguiculata.htm [Accessed: December 10, 2022]

[22] Britannica, The Editors of Encyclopaedia. "cowpea". *Encyclopedia Britannica*. 2022. Available from: https://www.britannica.com/plant/cowpea [Accessed: December 15, 2022]

[23] Weng Y, Ravelombola W, Yang W, Qin J, Zhou W, Wang Y, et al. Screening of seed soluble sugar content in cowpea (*Vigna unguiculata* (L.) Walp). American Journal of Plant Sciences. 2018;**9**:1455-1466. DOI: 10.4236/ajps.2018.97106

[24] Horn LN, Shimelis H. Production constraints and breeding approaches for cowpea improvement for drought prone agro-ecologies in Sub-Saharan Africa. Africa Annals of Agricultural Sciences. 2020;**65**(1):83-91. DOI: 10.1016/j.aoas.2020.03.002

[25] Narayana Health. Available from: https://www.narayanahealth.org/blog/health-benefit-of-cowpeas [Accessed: December 13, 2022]

[26] Abebe BKAND, Alemayehu MT. A review of the nutritional use of cowpea (*Vigna unguiculata* L. Walp) for human and animal diets. Journal of Agriculture and Food Research. 2022;**2022**:10. DOI: 10.1016/j.jafr.2022.100383

[27] Majola NG, Gerrano AS, Amelework A, Shimelis H, Swanevelder D. Genetic diversity and population structure analyses of South African Bambara groundnut (*Vigna subterranea* [L]. Verdc.) collections using SNP markers. South African Journal of Botany. 2022;**150**:1061-1068. DOI: 10.1016/j.sajb.2022.09.008

[28] Available from: https://greenwaymyanmar.com/posts/

pests-of-groundnut-and-control-measures [Accessed December 10, 2022]

[29] Department of Agriculture, Forestry and Fisheries. Bambara Groundnuts (*Vigna subterranea*). 2016. Available from: https://www.dalrrd.gov.za/Portals/0/Brochures%20and%20Production%20guidelines/Bambara.pdf [Accessed: December 15, 2022]

[30] Uba CU, Oselebe HO, Tesfaye AA, Agbo CU, Abtew WG. Comparative analysis of variation in African Bambara groundnut [*Vigna subterranea* (L) Verdc.] landraces assessed through seed traits. South African Journal of Botany. 2022;**150**:48-57. DOI: 10.1016/j.sajb.2022.06.060

[31] Khan MMH, Rafii MY, Ramlee SI, Jusoh M, al Mamun H. Bambara Groundnut (*Vigna subterranea* L. Verdc): A crop for the new millennium, its genetic diversity, and improvements to mitigate future food and nutritional challenges. Sustainability. 2021;**2021**:13. DOI: 10.3390/su13105530

[32] Tan XL, Azam-Ali S, Goh EV, Mustafa M, Chai HH, Ho WK, et al. Bambara groundnut: An underutilized leguminous crop for global food security and nutrition. Frontiers in Nutrition. 2020;7:601496. DOI: 10.3389/fnut.2020.601496

[33] Van Wyk BE. The potential of South African plants in the development of new food and beverage products. African Journal of Botany. 2011;77:857-868. DOI: 10.1016/j.sajb.2011.08.003

[34] Shayanowako AIT, Morrissey O, Tanzi A, Muchuweti M, Mendiondo GM, Mayes S, et al. African leafy vegetables for improved human nutrition and food system resilience in southern Africa: A scoping review. Sustainability. 2021;**13**(5):2896. DOI: 10.3390/su13052896

[35] Kongdang P, Dukaew N, PruksakornD,KoonrungsesomboonAND, N. Biochemistry of *Amaranthus* polyphenols and their potential benefits on gut ecosystem: A comprehensive review of the literature. Journal of Ethnopharmacology. 2021;**5**:281. DOI: 10.1016/j.jep.2021.114547

[36] Gerrano AS, Jansen van Rensburg WS, Adebola PO. Genetic diversity of *Amaranthus* species in South Africa. South African Journal of Plant and Soil. 2015;**32**(1):39-46. DOI: 10.1080/02571862.2014.973069

[37] Adegbola PI, Adetutu A, Olayini TD. Antioxidant activity of Amaranthus species from the Amaranthaceae family – A review. South African Journal of Botany. 2020, 2020;**133**:111-117. DOI: 10.1016/j.sajb.2020.07.003

[38] Xuan TD, Khanh TD. Chemistry and pharmacology of *Bidens pilosa*: An overview. Journal of Pharmaceutical Investigation. 2016;**46**:91-132. DOI: 10.1007/s40005-016-0231-6

[39] Yildirim H, Özdöl T, Yaşayacak H. An alien species of Bidens (Asteraceae): *Bidens pilosa* L., new to the Turkish flora. Acta Biologica Turcica. 2022;**32**(1):61-64.http://www.actabiologicaturcica.com

[40] Bartolome AP, Villaseñor IM, Yang W-C. *Bidens pilosa* L. (Asteraceae): Botanical properties, traditional uses, phytochemistry, and pharmacology. Evidence-Based Complementary and Alternative Medicine. 2013;**2013**:1549-1575. DOI: 10.1155/2013/340215

[41] Mtenga DV, Ripanda AS. A review on the potential of underutilized Blackjack *(Biden Pilosa)* naturally occurring in sub-Saharan Africa. International Journal of Food Properties. 2022;**25**(1):1549-1575. DOI: 10.1080/10942912.2022.2064491© 2022

[42] Singh G, Passsari AJ, Singh P, Leo VV, Subbarayan S, Kumar B, et al. Pharmacological potential of Bidens Pilosa L. and determination of bioactive compounds using UHPLC-QqQLIT-MS/MS and GC/MS. BMC Complementary and Alternative Medicine. 2017;**17**:492. DOI: 10.1186/s12906-017-2000-0

[43] Mashamaite CV, Manyevere AAND, Chakauya E. *Cleome gynandra*: A wonder climate-smart plant for nutritional security for millions in semi-arid areas. Frontiers in Plant Science. 2022;**13**:1003080. DOI: 10.3389/fpls.2022.1003080

[44] Van den Heever EAND, Venter SL. Nutritional and medicinal properties of *Cleome gynandra*. Acta. Horticulture. 2007;**752**:127-130. DOI: 10.17660/ActaHortic.2007.752.17

[45] Aremu AO, Moyo M. Health benefits and biological activities of spiny monkey orange (*Strychnos spinosa* Lam.): An African indigenous fruit tree.2022.283.114704. Journal of Ethnopharmacology. 2022:283. DOI: 10.1016/j.jep.2021.114704

[46] Nethavhani M. *Cochorus olitorius.*pza.sanbi.org/Corchorus-olitorius [Accessed: January 04, 23]

[47] Gbèdoloa AE, Dassou AG, Dassou HG, Aminon ID, Omondi BA, Dansi A. Morphotype diversity of *Corchorus olitorius* and influence of agricultural practices on its potential major pest insects. Scientia Horticulturae. 2018;**239**(15):234-241. DOI: 10.1016/j.scienta.2018.05.047

[48] Ndhlovu J, Afoloyan AJ. Nutritional analysis of South African wild vegetable *Corchorus olitorius* L. Asian Journal of Plant Sciences. 2008;**7**(6):615-618. DOI: 10.3923/ajps.2008.615.618

[49] Pereira PR, Silva JT, Verícimo MA, Paschoalin VMF, Gerlinde Teixeira GAPB. Crude extract from taro (Colocasia esculenta) as a natural source of bioactive proteins able to stimulate haematopoietic cells in two murine models. Journal of Functional Foods. 2015;**18**:333-343. DOI: 10.1016/j.jff.2015.07.014

[50] Mitharwal S, Kumar A, Chauhan K, Taneja NK. Nutritional, phytochemical composition and potential health benefits of taro (*Colocasia esculenta* L.) leaves: A review. Food Chemistry. 2022;**383**:132406. DOI: 10.1016/j.foodchem.2022.132406

[51] Prajapati R, Kalariya M, Umbarkar R, Parmar S, Sheth N. Colocasia esculenta: A potent indigenous plant. International Journal of Nutrition, Pharmacology, Neurological Diseases. 2011;**1**:2. DOI: 10.4103/2231-0738.84188

[52] Pereira PR, de Aquino Mattos EB, Corrêa NACF, Vericimo MA, Paschoalin VMF. Anticancer and immunomodulatory benefits of Taro (*Colocasia esculenta*) corms, an underexploited tuber crop. International Journal of Molecular *Science*. 2021;**22**(1):265. DOI: 10.3390/ijms22010265

[53] Pawar HA, Choudhary PD, Kamat SR. An overview of traditionally used herb, *Colocasia esculenta*. Phytomedicine. 2021;**218**:4. DOI: 10.4172/2167-0412.1000317

[54] Pfukwa TM, Chikwanha OC, Chenaimoyo LF, Katiyatiya CLF, Fawole OA, Manleya M, et al. Southern African indigenous fruits and their byproducts: Prospects as food antioxidants. Journal of Functional Foods. 2020;**2020**(75):104220. DOI: 10.1016/j.jff.2020.104220

[55] Mashau ME, Kgatla TE, Makhado MV, Mikasi MS, Ramashia SE.

Nutritional composition, polyphenolic compounds and biological activities of marula fruit (*Sclerocarya birrea*) with its potential food applications: A review. International Journal of Food Properties. 2022;**8**(6):1549-1575. DOI 10.1080/10942912.2022.2064491

[56] Available from: https://ar.pinterest.com/pin/386605949240381857 [Accessed: December 10, 22]

[57] Mariod AA, Abdelwahab SI. *Sclerocarya birrea* (Marula): An African Tree of Nutritional and Medicinal Uses: A Review. Food Reviews International. 2012;**28**(4):375-388. DOI: 10.1080/87559129.2012.660716

[58] Nitcheu Ngemakwe PH, Remize F, Thaoge M, Sivakumar D. Phytochemical and nutritional properties of underutilised fruits in the southern African region. South African Journal of Botany. 2017;**113**:137-149. DOI: 10.1016/j.sajb.2017.08.006

[59] Omotayo AO, Aremu AO. Undervalued Spiny Monkey Orange (*Strychnos spinosa* lam.): An indigenous fruit for sustainable food-nutrition and economic prosperity. Plants (Basel). 2021;**10**:2785. DOI: 10.3390/plants10122785

[60] Ngadze RT, Linnemann AR, Nyanga LK, et al. Local processing and nutritional composition of indigenous fruits: The case of monkey orange (*Strychnos* spp.) from Southern Africa. Food Review International. 2016;**33**(2):123-142. DOI: 10.1080/87559129.2016.1149862

[61] Akweni AL, Sibanda S, Zharare GE, Zimudzi C. Fruit-based allometry of *Strychnos madagascariensis* and *S. spinosa* (Loganiaceae) in the savannah woodlands of the Umhlabuyalingana municipality, KwaZulu-Natal, South Africa. Trees, Forests and People. 2020;**2020**:2. DOI: 10.1016/j.tfp.2020.100025

[62] Rocha MI, Rodrigues MJ, Pereira C, Pereira H, da Silva MM, da Rosa NN, et al. Biochemical profile and in vitro neuroprotective properties of *Carpobrotus edulis* L., a medicinal and edible halophyte native to the coast of South Africa. South African Journal of Botany. 2017;**111**:222-232. DOI: 10.1016/j.sajb.2017.03.036

[63] Kolosa L. *Carpobrotus edulis*: Invasive food and medicine. Available from: https://lindsaykolasa.com/2020/08/01/carpobrotus-edulis-invasive-food-medicine/ [Accessed: December 13, 2022]

[64] Akinyede KA, Ekpo OE, Oguntibeju OO. Ethnopharmacology, therapeutic properties and nutritional potentials of *Carpobrotus edulis*: A comprehensive review. Scientia Pharmaceutic. 2020;**88**:39. DOI: 10.3390/scipharm88030039

[65] Van der Watt E, Pretorius CJ. Purification and identification of active antibacterial components in *Carpobrotus edulis* L. Journal of Ethnopharmacology. 2001;**76**:87-91. DOI: 10.1016/s0378-8741(01)00197-0

[66] Hafsa J, Hammi KM, Le Cerf D, Limem K, Majdoub H, Charfeddine B. Characterization, antioxidant and antiglycation properties of polysaccharides extracted from the medicinal halophyte Carpobrotus edulis L. International Journal of Biological Macromolecules. 2018;**107**:833-842. DOI: 10.1016/j.ijbiomac.2017.09.046